PESTICIDE POLICY, PRODUCTION RISK, AND PRODUCER WELFARE

AN ECONOMETRIC APPROACH TO APPLIED WELFARE ECONOMICS

JOHN M. ANTLE

A STUDY FROM RESOURCES FOR THE FUTURE

Washington, D.C.

Printed in the United States of America

Published by Resources for the Future
1616 P Street, N.W., Washington, D.C. 20036
Books from Resources for the Future are distributed worldwide by The Johns Hopkins University Press

Library of Congress Cataloging-in-Publication Data

Antle, John M.
Pesticide policy, production risk, and producer welfare: an econometric approach to applied welfare economics / by John M. Antle.
p. cm.
Bibliography: p.
ISBN 0-915707-39-X
1. Agricultural productivity—United States. 2. Pesticides—Economic aspects—United States. I. Resources for the Future. II. Title.
HD1765.A58 1988
363.7'384—dc19 88-14120
CIP

The book was edited by Samuel Allen and designed by Debra Naylor.

∞ The paper in this book meets the guidelines for permanence and durability of the Committee on Production Guidelines for Book Longevity of the Council on Library Resources.

RESOURCES FOR THE FUTURE

RESOURCES FOR THE FUTURE (RFF) is an independent nonprofit organization that advances research and public education in the development, conservation, and use of natural resources and in the quality of the environment. Established in 1952 with the cooperation of the Ford Foundation, it is supported by an endowment and by grants from foundations, government agencies, and corporations. Grants are accepted on the condition that RFF is solely responsible for the conduct of its research and the dissemination of its work to the public. The organization does not perform proprietary research.

RFF research is primarily social scientific, especially economic. It is concerned with the relationship of people to the natural environmental resources of land, water, and air; with the products and services derived from these basic resources; and with the effects of production and consumption on environmental quality and on human health and well-being. Grouped into five units—the Energy and Materials Division, the Quality of the Environment Division, the Renewable Resources Division, the National Center for Food and Agricultural Policy, and the Center for Risk Management—staff members pursue a wide variety of interests, including forest economics, natural gas policy, multiple use of public lands, mineral economics, air and water pollution, energy and national security, hazardous wastes, the economics of outer space, and climate resources. Resident staff members conduct most of the organization's work; a few others carry out research elsewhere under grants from RFF.

Resources for the Future takes responsibility for the selection of subjects for study and for the appointment of fellows, as well as for their freedom of inquiry. The views of RFF staff members and the interpretation and conclusions of RFF publications should not be attributed to Resources for the Future, its directors, or its officers. As an organization, RFF does not take positions on laws, policies, or events, nor does it lobby.

CONTENTS

LIST OF TABLES

LIST OF FIGURES

ACKNOWLEDGMENTS

This study was initiated while I was a Gilbert White Fellow at Resources for the Future (RFF) in 1985, on sabbatic leave from the University of California, Davis. It was completed in 1986 while I was a Resident Fellow with the National Center for Food and Agricultural Policy at RFF and a visiting professor at the University of Maryland. The generous support of each of these organizations, both direct and indirect, made the completion of this study possible, and I thank each of them.

My interest in pest management and pesticide policy grew out of my involvement in the research funded by the Statewide Integrated Pest Management (IPM) Program at the University of California. That program funded a project for the collection of data on and an economic analysis of the tomato IPM program. The case study in chapter 5 is based on the data collected for that project. Thanks go to James Lyons, director of the University of California IPM program, and to Mary Louise Flint, Karen Klonsky, Theodore Wilson, and Frank Zalom, who provided valuable advice during the early stages of the original study of the tomato IPM program. The data survey was designed and carried out with the most able assistance of Seong Park, now an economist at the Korea Rural Economics Institute, to whom I also owe many thanks. His Ph.D. dissertation provided much of the background and preliminary analysis for the case study.

I also owe thanks to many individuals at Resources for the Future. John F. Ahearne's suggestions at an early stage in the writing were most helpful in achieving a balance in the manuscript between breadth and depth. Susan M. Capalbo, Kenneth D. Frederick, Michael Hazilla, Allen V. Kneese, Raymond J. Kopp, Alan J. Krupnick, and Tim T. Phipps provided comments on various parts of the material contained in this manuscript at various stages in its development. The productive, collegial environment at RFF contributed much to the eventual completion of the manuscript. Finally, I wish to thank the three peer reviewers for their thoughtful suggestions, which led to improvements throughout the manuscript, and Samuel Allen, who skillfully edited the manuscript.

one

INTRODUCTION

Production risk is the farmer's perennial problem, and agricultural pests are a major source of production risk. Since earliest times farmers have used a variety of pest management techniques to reduce the adverse effects of pests. Until the twentieth century, pest management practices were largely biological in nature, involving the use of crop rotations, timing of planting, and other techniques that inhibited the growth of pest populations. The discovery during the 1940s of the pesticidal properties of dichloro-diphenyl-trichloroethane (DDT) and other chlorinated hydrocarbons, and of organophosphates such as parathion, led to their rapid dissemination in the postwar era. Since the 1950s many other chemical materials have been developed for use by farmers in their ongoing attempt to insure themselves against the risk of pest damage to their crops.

Pesticides are essentially poisons, however, and their widespread use in agriculture creates difficult problems for producers and for society at large. Despite the apparent benefits of chemical pesticides to farmers, there has been growing public concern that agricultural pesticides are a major source of pollution which has detrimental effects on human health and the natural environment. Pesticide use is also a controversial issue within agriculture. While evidence suggests that agricultural pesticides have made a substantial contribution to the growth of productivity in the second half of this century, phenomena such as pest resistance and the elimination of natural enemies have raised serious questions about the value of chemical-intensive pest management technology. Some agricultural scientists question whether a pest management technology of this kind which disrupts the ecosystem can sustain the high levels of productivity achieved since the 1950s.

The available data suggest that the use of chemical pesticides, and the problems associated with them, will not diminish in the near future. The United States Department of Agriculture (USDA) estimates that in 1982 about 450 million pounds of active ingredient of herbicides were used in U.S. agriculture, as compared to some 207 million pounds in 1971. In the case of insecticides, about 71 million pounds were used in 1982, compared to 126 million pounds in 1971. While these numbers mask important differences in the types of chemicals applied, it is clear that large quantities of these materials have been, are being, and will continue to be applied in U.S. agriculture.

Agricultural pesticides pose a dilemma for agricultural producers and society at large because, on the one hand, pesticides appear to have beneficial effects on the quantity and quality of agricultural production, but on the other hand may impose costs on both producers and society. Economic analysis demonstrates that the policy of laissez faire is not in the public's interest because of the likelihood that markets will fail to take into account all of the benefits and costs to society that are associated with pesticides. Existing public policies recognize the need for government intervention to deal with the pesticide problem. The Federal Insecticide, Fungicide, and Rodenticide Act (FIFRA), enacted in 1947, and subsequent amendments to it empower the Environmental Protection Agency (EPA) to evaluate the effects of pesticides on human health and the environment, and to regulate pesticide use as needed to balance social benefits and costs. Many state governments have undertaken similar regulatory responsibilities. Thus, government regulation of pesticide use by both federal and state agencies will be a part of public policy for the forseeable future.

Given society's decision to endow government with the responsibility to deal with these issues, the question arises as to what policies should be implemented. There are two dimensions to existing pesticide policy. The first involves determining which materials pose a threat to human health or the environment, and restricting or prohibiting the use of those materials. Economists have a general criterion for determining when regulation is required: each pesticide should be used in such a way that the positive benefits, summed over all members of society (including consumers of agricultural products and agricultural producers) should exceed the costs to all members of society (including the direct, or "private," resource costs of producing and applying the pesticides that are paid by agricultural producers, and the indirect, or "social," costs of pest resistance, pollution, and human health risks). From this perspective, the government's role should be to determine the social benefits and costs associated with the use of agricultural pesticides, and where unregulated use of a pesticide fails to yield net benefits to society, to design policies that correct the problem.

While this regulatory scheme is straightforward in principle, it is fraught with many problems in practice. One of the most formidable is obtaining all of the information needed to make a regulatory decision: knowledge of the benefits and costs to agricultural producers and society at large of the thousands of agricultural pesticides in existence. It is especially difficult to determine the external costs of pesticides—their effects on pest resistance, the environment, and human health. Some of these effects can in principle be quantified; others involve valuation of nonmarket goods; still others, such as the valuation of human life, are highly controversial. The scientific foundations and data for quantifying the external effects of pesticides range from very solid to quite shaky. In the case of pest resistance to certain chemicals, there is sound scientific understanding and ample data. Also, effects of certain pesticides on human health are well known. But in many cases the basic science is not well understood, nor is the epidemiological evidence adequate for drawing reliable inferences about health effects. There are equally difficult problems in attempting to measure the broader environmental effects, since the movement of pesticides through the environment and their effects on it are not well understood. (For further discussion of these issues, see Antle and Capalbo [1986].)

The second and longer-run dimension of pesticide policy concerns the need for the public sector to invest in scientific research and development that will produce technological alternatives to agricultural pesticides—alternatives that are economically viable for producers and are more compatible with human health and the environment. Both federal and state governments have devoted resources to such research, generally under the rubric of "integrated pest management," often referred to as IPM. IPM strives to consider the entire spectrum of pest control options, such as biological control, genetic engineering, and cultural practices, with chemical pesticides being considered for use only when they are the most effective means of control.

The purpose of this study is to provide an econometric framework for the measurement and analysis of the direct economic benefits that agricultural producers derive from the use of agricultural pesticides and other pest management practices such as IPM, and to illustrate how that framework can be used by examining its application in a case study of California processing-tomato production. In view of the difficulty of determining the social costs associated with pesticides, it seems reasonable to suggest that analysis of the pesticide problem should begin with the evaluation of the private benefits and costs to the producers who use them. If a pesticide does not yield net benefits to the agricultural producers themselves, economically efficient farmers will not use the pesticide, and its regulation will affect only inefficient producers and will necessarily increase social welfare. If a pesticide does yield net benefits to producers, and producers use the pesticide efficiently, then a binding regulation such as a restriction on use generally will have an adverse effect on the economic welfare of producers. Thus when there appears to be a need for regulation as a result of pollution and other harmful side effects of a pesticide, policy makers need to know the magnitude of net benefits to producers from use of the pesticide as they attempt to balance those private net benefits against the social costs. Moreover, if policies are to be designed to produce technological alternatives to pesticides and to encourage farmers to use those alternatives, the economic and technological attributes of the pesticide-based technology and the alternative technologies must be understood as they are perceived by the farmers who use them.

The current practice at EPA and other regulatory agencies is to estimate the cost of pesticide restrictions in terms of the estimated yield losses. There are at least two problems with this procedure. First, such estimates tend to overestimate the yield losses whenever it is possible for farmers to make input or output substitutions and thus counteract the economic effects of the restrictions. Second, even if the yield loss estimates are accurate, they represent the effect of the policy on the farmer's welfare after production has taken place—that is, they measure the farmer's ex post welfare. The yield loss estimates therefore fail to account for the effects of the policy on the farmer's welfare when production decisions must be made—that is, on the farmer's ex ante welfare. According to economic theory, it is the effects of a policy on ex ante welfare that determines economic behavior such as pest management decisions. Thus, this study is concerned with the effects of policy on the farmer's ex ante welfare when pest management decisions are being made.

To evaluate ex ante welfare associated with pest management decisions, it must be recognized that pest management is characterized by two properties that differentiate it from many other production activities, and that as a result conventional economic analysis of the effects of pesticides on producer welfare is not appropriate and may be misleading. First, the productivity of pest management inputs typically depends on a random event, namely the presence of the pest. If the pest occurs in the crop, the pest management input has an effect on production; otherwise it does not, except for possible secondary effects on beneficial organisms or on plant growth. In other words, pest management productivity is integrally related to the concept of *production risk*. Second, because the productivity of pest management depends on the pest population, pest management schemes often involve the acquisition over time of information about the pest population. The time dimension enters pest management in two ways: through intraseasonal decisions (on the application of pesticides during the growing season of a crop, for instance), and through interseasonal decisions (on crop rotation to control soil diseases and pests, for example). The former involve the sequential acquisition of information during the growing season in relation to the biological growth of the crop. The latter involve acquisition of information on biological interrelationships across growing seasons—information having to do with such phenomena as pest life cycles and pest resistance. Thus pest management productivity—and its measurement—also is integrally related to the dynamics of the production process.

In order to measure the ex ante value of pesticides as perceived by farmers, it is necessary to measure the economic attributes of the technology and how farmers value those attributes. And since one of the important dimensions of pesticide productivity is hypothesized to be risk-related, the empirical framework for the evaluation of producer welfare must involve *measurement of stochastic (risky) technologies* and *measurement of producers' risk attitudes*.

The conventional framework economists use to analyze producer behavior—known as the neoclassical theory of production—is both timeless (static) and certain (riskless). This theory is not useful, however, when dynamics and risk are essential elements of the production problem. The production theory and econometric methodology proposed in the present study are dynamic, stochastic generalizations of the neoclassical theory and conventional econometric methods, and are valid for analysis of pesticide productivity or any other production problem in which dynamics and risk are important.

In attempting to generalize the neoclassical theory to account for uncertainty, one must confront several theoretical problems that arise in welfare economics. The conventional approach to welfare measurement is to use money equivalent measures of welfare, derived from areas under demand or supply curves. Under certainty, there are well-known difficulties that arise in measuring welfare in this way. For example, the estimated supply or demand functions must satisfy certain theoretical properties. Under uncertainty, additional problems arise: the sources of uncertainty must be quantified, and their effects on supply or demand curves must be taken into account. All these considerations make applied welfare analysis a difficult undertaking.

The alternative approach to welfare measurement pursued in this study is to estimate the risk attributes of the production technology and the risk attitudes of the producer population, and use them to construct an approximation to the ex ante welfare function of the decision maker. Unfortunately, it is not possible to know a priori what an individual's welfare function is, and applied welfare analysis must be based on some assumed function. In this study, the decision maker's welfare is defined in terms of expected utility. Based on an approximate expected utility function, welfare changes can be computed in terms of utility or transformed into a money equivalent (or money metric) measure of welfare.

As the title of the present study suggests, the framework proposed here is based on econometric analysis and thus is necessarily written for a rather specialized audience. Yet the issue of pesticide productivity is of interest to a much larger audience. It is important for all those involved in the public debate on environmental policy to be aware of the conceptual and methodological foundations of research aiming to quantify the private and social benefits and costs of pesticides. For this reason, chapter 2 and the case study presented in chapter 5 are written so that the general economist or the informed lay person can understand the conceptual issues involved. The two intervening chapters are written for the research audience, and necessarily delve into the theoretical and methodological details underlying the implementation of the general framework being proposed.

It is hoped that this study will make contributions on several fronts. The measurement of producer welfare under uncertainty has been the topic of recent theoretical research, but little progress has been made in quantifying the relevant relationships in a way that is useful for policy analysis. Thus, one goal of this study is to develop an empirical framework in which producer welfare under uncertainty can be analyzed. The quantification of welfare analysis is important on two levels. On the one hand, theoretical research has raised questions about the validity of conventional welfare analysis when production uncertainty is important. Using the methods proposed here, it is possible to investigate quantitatively the degree to which risk can bias conventional welfare analysis. On the other hand, for welfare economics to be useful to policy analysts, economists must be able to quantify the relevant variables.

While the concepts and methods discussed are quite general and could be applied to a wide range of problems, I have chosen to focus on the pesticide question for two reasons. First, pest management provides perhaps the best example of a production process of which both dynamics and uncertainty are integral parts. Second, I believe the question of pesticide regulation is and will continue to be a major focus of public policies dealing with agriculture. I foresee one of the principal agricultural policy issues in the future to be the regulation of agriculture. One of the reasons for such regulation will be public concern about the environmental and health effects of pesticides.

The central theoretical and methodological chapters of the study—chapters 3 and 4—reflect the current state of my efforts to develop econometric methods suitable for the measurement of the technologies and behavioral attributes of a producer population. Using these methods, I believe it is possible to learn more about the role that production risk plays in producer behavior than was possible with the methods previously available. This study represents an ongoing effort to integrate the essential elements of production risk, sequential decision making, and attitudes toward risk into a coherent analytical and empirical framework. I believe that progress has been made toward the goal of a unified approach, but as the concluding chapter indicates, some important elements of a general methodology remain elusive.

It is also hoped that this study will contribute, through the case study presented in chapter 5, to an understanding of the roles that alternative methods of pest management—both use of pesticides and integrated management techniques—play in producer welfare. Agricultural scientists have long understood the potential problems associated with agricultural pesticides, and have consequently advocated a variety of pest management techniques—those techniques referred to as integrated pest management. Advocates of IPM stress that there are often ecologically and economically sound alternatives to purely chemical-based pest management strategies. The findings of this study, I believe, provide an additional, powerful argument in favor of IPM as a policy tool. I find in the case study of processing-tomato production presented in chapter 5 that pesticide restrictions impose greater welfare costs on those farmers who face more risk and who are more risk averse. At the same time, I find that an IPM program which increases the farmer's ability to effectively use pesticides yields the greatest benefits to the very farmers who are most adversely affected by the restrictions. To the extent that they prove to be generally valid, these findings suggest that IPM programs which substitute for the risk-reducing effects of pesticides may offset the welfare costs of pesticide regulations in an equitable manner, by providing the most benefits to those farmers most adversely affected by the regulations.

two

AN OVERVIEW OF THE ISSUES

This chapter begins with background information about pesticides, pest management, and pesticide policy, then introduces the reader to the theoretical and econometric issues that are addressed in detail in chapters 3 and 4. Following a discussion of the conventional approach to the analysis of firm welfare, as based on the neoclassical theory of production, the text examines how the neoclassical theory must be generalized when dynamics and uncertainty are an important part of the production process.

Pesticides and Pest Management

An agricultural pest is an organism whose presence may have undesirable effects on the production of an agricultural product. The most common agricultural pests are other plants (weeds), insects, invertebrate and vertebrate animals, and microorganisms (such as bacteria and fungi). The effects of a pest may be measured in a variety of ways; most frequently it is done in terms of crop yield or crop quality. It will be argued here that agricultural pests have an impact not only on average yield and quality, but also on the degree of *risk* associated with yield or crop quality.

The goals and objectives of pest management have been, from earliest times, to reduce the adverse effects of agricultural pests. By reducing the yield losses caused by these pests, farmers can enhance the returns to the other resources devoted to production. Early pest control techniques included hand removal of plant and insect pests, and modification of the environment by burning fields and rotating crops so that it was less favorable to pests. Biological control, such as the introduction of pests' natural enemies, has a history dating back to ancient times. Chemicals like arsenic have been used for several millennia. Until the late nineteenth and the twentieth centuries, however, chemical control was neither widespread nor particularly successful.

The farmer's ability to control and even eliminate many pests, especially insects, seemed to take a great leap forward in the 1940s and 1950s with the discovery of DDT's properties. This and other chlorinated hydrocarbons, organophosphates, and carbamates provided the basis for the series of "miracle" insecticides, herbicides, and fungicides that were rapidly developed and adopted for use in agricultural production in the decades following World War II. During the 1950s and 1960s it was commonly believed that widespread use of chemical pesticides could solve agricultural pest problems at a low cost to society.

The first indication that the miracle pesticides had their limitations came in the 1950s when entomologists found that some insects developed resistant strains after being treated with certain chemicals. In addition, farmers began to find that previously benign insects, whose populations had been held in check by natural enemies, became major pests when the new pesticides killed their natural enemies. It also became increasingly evident throughout the 1960s that pesticides were having harmful effects on human health and on the natural environment.

The harmful side effects of agricultural pesticides, or *externalities*, are the result of several factors. First of all, pesticides are essentially poisons (indeed, organophosphates were originally developed during the 1940s by German scientists for use in war as a nerve gas). It is not surprising that materials that are effective in killing certain plant or insect pests also may harm other plants, insects, mammals, and fish. Second, the ecosystem of the pest and other organisms is disrupted by the introduction of these poisons on a large scale. Chemically induced disruptions of the ecosystem can have widely differing effects, depending on a variety of factors. For example, if a chemical kills many kinds of insects, it is likely to kill the pest insect as well as other insects with which the pest competes in the ecosystem. One frequently observed phenomenon when such broad-spectrum insecticides are used is an initial decline of the pest population, followed by resurgence of the pest insect because of the reduction of its natural enemies and predators. This problem may be especially severe if the pest acquires genetic resistance to the pesticide.

The origins of integrated pest management lie in the recognition by entomologists and other scientists, beginning in the 1950s, that the widespread use of chemicals to control agricultural pests may disrupt the ecosystem and thus fail to provide a long-term pest control solution. Thus, at the heart of the entomologist's concept of IPM is the relation of pest management to the ecosystem. Flint and van den Bosch (1981) provide the following definition of integrated pest management:

> IPM is an ecologically based pest management control strategy that relies heavily on natural mortality factors such as natural enemies and weather and seeks out control tactics that disrupt these factors as little as possible. IPM uses pesticides, but only after systematic monitoring of pest populations and natural control factors indicates a need. Ideally, an integrated pest management program considers all available pest control actions, including no action, and evaluates the potential interaction among various control tactics, cultural practices, weather, other pests, and the crop to be protected.

More specifically, Flint and van den Bosch outline six basic elements of IPM:

> (1) people: the system devisers and pest managers; (2) the knowledge and information necessary to devise the system and make sound management decisions; (3) a program for monitoring the numbers and state of the ecosystem elements—e.g., resource, pest and natural enemies; (4) decision-making levels: the pest densities at which control methods are put into action; (5) IPM methods: the techniques used to manipulate pest populations; and (6) agents and materials: the tools of manipulation.

Flint and van den Bosch emphasize that "integrated pest management systems are dynamic, as are the ecosystems in which they are invoked, and usually involve continuous information gathering and evaluation as the resource and its associated physical and biological environment go through their seasonal progressions." Thus, according to entomologists, integrated pest management involves sequential pest management decisions. For an economic analysis of pest management, then, it is important that economic models of the production process account for the sequential nature of pest management decisions.

An important issue in the analysis of pest management is *producer behavior*. How do farmers use pesticides, and how should they? What constitutes "efficient" pesticide use? Agricultural economists have found substantial evidence that farmers operate, as most private businesses do, to obtain as high a return as possible on the resources they have invested. This evidence suggests that farmers try to use pesticides in an economically efficient way, as they do other production inputs. But there seems to be a consensus among entomologists that farmers have not taken into account the ecological effects of pesticides, so that from the ecological point of view there may have been widespread overuse of certain chemicals in agriculture. Environmental economists agree, arguing that farmers ignore the effects of pesticides on human health and on the environment outside agriculture and therefore overuse pesticides.

To reconcile these interpretations of how pesticides should be used, three different concepts of efficiency can be identified; they are associated with the points of view of the farmer, the entomologist, and the economist. Efficiency, from the farmer's point of view, takes into account the perceived benefits and costs of the pesticides. However, farmers can only make decisions about pesticide use in terms of their knowledge of the pesticides and related pest management techniques. When many new chemical pesticides were being introduced in the 1950s and 1960s, relatively little was known about either the direct effects of pesticides on the farm or the harmful side effects of the chemicals. Thus it is likely that farmers did misuse or overuse pesticides. But as farmers became better informed about pest management based on the new chemicals, it seems unlikely that they continued to use them inefficiently in terms of the perceived benefits and costs on the farm.

Efficient pesticide use from the farmer's point of view involves the maximization of economic returns to the resources invested in the farm enterprise. This idea is embodied in the concept of the "economic threshold" for pesticide use (see Headley, 1972). The economic threshold is the level of pest population at which the private benefits of pesticide use outweigh the costs. If farmers who apply a pesticide according to a schedule or when a pest first appears are likely to overuse the pesticide, application according to an economic threshold would increase efficiency and help reduce overuse of pesticides.

Entomologists have found that pesticides have many ecological effects that farmers may not take into account in their pesticide decisions. Thus, even if application of an economic threshold reduced inefficient pesticide use in terms of private costs and benefits, it would not necessarily reduce the adverse ecological effects of pesticides. Entomologists could in principle develop alternative "ecological" thresholds for pesticide use which would take into account the broader on-site and off-site effects of pesticides, but such decision rules would likely conflict with the farmer's economic efficiency criterion.

Economists studying environmental issues have observed that if the use of a chemical pesticide imposes external costs on the agricultural sector or on other members of society, farmers may overuse pesticides even if they are applied according to an economic or ecological threshold. Although a biological scientist might design decision rules for pesticide use that could account for the ecological effects of pesticides, such a threshold would not necessarily take into account the costs that pesticides may impose on other members of society. For example, if pesticide runoff from fields results in water pollution, the farmer's use of the pesticide, in addition to other ecological effects, imposes a cost on society if society must purify the drinking water. The cost of water purification is not taken into account in the farmer's decision to use the pesticide, or in the ecologist's evaluation of the pesticide's environmental effects. If the cost of the water pollution were taken into account in the farmer's decision to use a pesticide—through the imposition of a tax on the pesticide, for instance—less pesticide would be used. From society's point of view, then, a farmer may be overusing a pesticide that causes water pollution even if it is being used efficiently from the farmer's point of view or from the ecologist's point of view.

Another basic concept is *production technology*. Economists define a technology as a "way of producing," or as a set of production practices. Farmers choose among production practices in deciding how to produce. When a specific subset of these practices is implemented, the resulting relationship between inputs and outputs is defined as the *production function*. In the neoclassical theory, described below, the relationship between inputs and outputs is deterministic; a given input results in a specific output. More generally, there may be an element of uncertainty involved in the link between inputs and outputs. Such *production uncertainty* is a result of the effects of weather, pests, and other random factors beyond the control of the farm manager.

Various types of inputs, such as land, human labor, machinery, seeds, fertilizers, and pesticides generate service flows into the production process. Each set of inputs in a farm operation generates a service flow: tractors, plows, and labor are used to produce service flows for field preparation; other inputs generate service flows for planting, cultivation, pest management, irrigation, harvesting, and so on.

Input substitution is an important property of production technologies and production functions. Input substitution is typically possible between inputs at various levels in the production process. When a farmer makes capital investment decisions, a choice is being made between various possible combinations of capital and labor, for example. Several tractor operators can cultivate a given number of acres in a day with small tractors, or one operator can do so with a large tractor. Thus mechanical power in the form of a tractor can substitute for human labor. To take an example from pest management, herbicides can be used to control weeds in place of—as a substitute for—mechanical cultivation. The concept of input substitution plays an important role in the analysis of pest management, as it suggests that farmers may have technological alternatives to the use of chemical pesticides.

Pesticide Policy and Social Welfare

Throughout this century the federal and state governments have been involved to varying degrees in the regulation of the manufacture, trade, and use of pesticides. The laws passed in the early twentieth century were intended to protect consumers from foods contaminated by harmful materials; more recently, legislation has broadened to include protection of pesticide applicators, livestock, fish, and the natural environment in general. To understand the nature and scope of pesticide policy, it is useful to digress briefly on the history of pesticide laws and regulations at the federal level.

Pesticide Legislation

The first federal legislation dealing with pesticides was the Insecticide Act of 1910 (Public Law 61-152), which made it illegal to manufacture, sell, or transport adulterated or mislabeled materials. The Food, Drug, and Cosmetic Act of 1938 (P.L. 75-717) attempted to protect consumers from foods containing poisonous additives or residues. The introduction of new chemicals such as DDT after World War II made the earlier legislation obsolete. The Federal Insecticide, Fungicide, and Rodenticide Act of 1947 (P.L. 80-104), or FIFRA, which replaced the 1910 act, required that all pesticides must be registered with the U.S. Department of Agriculture (USDA) before their sale. Registration required that the pesticide label list all ingredients, specify appropriate uses, and bear instructions for use. The USDA administered these regulations, but did not enforce the regulations on actual pesticide use. Even if USDA refused registration of a pesticide, it could be sold under protest by the manufacturer; USDA was then required to prove that the pesticide was harmful to humans, animals, or vegetation.

The Pesticide Chemical Amendment of 1954 (P.L. 83-518) dealt further with the problem of residues of the new pesticides in foods. This act stipulated that USDA should not register a pesticide unless it was shown that harmful residues did not remain on agricultural products, or until an acceptable residue tolerance was set. The burden was on the manufacturer to show that the product was safe, and the Food and Drug Administration (FDA) could refuse proposed tolerances if the material was determined to be harmful to human health. During the 1950s concern was also growing about the environmental hazards posed by pesticides. The Pesticide Research Act of 1958 (P.L. 85-582) authorized the Department of the Interior to study the effects of pesticides on fish and wildlife.

The Food and Drug Law of 1958 (P.L. 85-791 and P.L. 85-929) added a particularly important new dimension to the testing and regulation of pesticides and other potential carcinogens. The section of the law known as the Delaney clause stated that "no additive shall be deemed to be safe if it is found to induce cancer when ingested by man or animal, or if it is found after tests which are appropriate for evaluation of safety of food additives to induce cancer in man or animal" (Coulston, 1979:xix). This clause generated heated controversy among toxicologists, biologists, and consumer advocates. Scientists have denounced it as a perversion of scientific principles, while environmentalists and consumer advocates have used it to justify increased regulation of pesticides and food additives.

Various amendments to FIFRA were enacted in the 1960s, but the most significant development was the release in the late 1960s of two congressional reports that were critical of USDA's registration procedures and of the Pesticide Regulation Divison within USDA. With the creation of the Environmental Protection Agency in 1970, authority for pesticide regulation (hitherto under USDA), for setting food tolerances (under the Department of Health, Education, and Welfare), and for research (under Interior) were all transferred to EPA. In the early 1970s EPA also took over the responsibility for establishing standards for protection of farm workers from pesticide hazards.

From the point of view of the regulating agencies, the principal weakness of FIFRA was that it did not provide any means of actually regulating the use of pesticides once they were registered. The Federal Environmental Pesticide Control Act of 1972 (P.L. 92-516), or FEPCA, replaced FIFRA by amendment. It went far beyond previous legislation, extending regulation over all pesticides produced, and covering both registration and use. Under FEPCA, the administrator of EPA was charged with publishing regulations and cooperating with other agencies to carry out the provisions of the act. Among the major provisions were the following:

1. *Environmental protection* was stated to be an explicit criterion in the regulation of pesticides. The act required that pesticides be regulated to avoid "unacceptable adverse effects on the environment," defined as "any unreasonable risk to man or the environment, taking into account the economic, social, and environmental costs and benefits of any pesticide."

2. *Registration* procedures were detailed requiring that manufacturers provide test data to support registration. Two classes of registration were created: general, for pesticides which required only

labeling; and restricted, for pesticides posing a health threat to applicators or others, and which could be used only by certified applicators. The latter could be further restricted by EPA regulations. These registration rules applied to materials manufactured or traded both intrastate and interstate.

3. *Cancellation and suspension* of registration were provided for. Registration would normally be cancelled after five years, but continued use would be allowed during reregistration. Under suspension, use would not be allowed. Suspension would allow the EPA administrator to immediately stop all sale and use of a pesticide when an "imminent hazard" was found. It is particularly noteworthy, for our purposes, that FEPCA stipulated that environmental protection should be based on a benefit-cost calculation, taking both private and social concerns into account.

As a result of the complexity and ambiguity of FEPCA, numerous controversies arose over its implementation. By late 1976 the overwhelming task of reregistering pesticides had brought the regulatory process at EPA to a virtual standstill. Among the major complaints was the claim that agricultural interests were being ignored. Consequently, a subsequent extension of FIFRA in 1975 (P.L. 94-140) required EPA to assess the effects of regulatory actions on the agricultural economy and to consult with the secretary of agriculture.

The Federal Pesticide Act of 1978 (P.L. 95-396) further modified the 1972 amendments to FIFRA and attempted to improve the implementation process. One of the most significant amendments in the 1978 law concerned administrative review for the cancellation of pesticides. The Rebuttable Presumption Against Registration (RPAR) process, outlined in the July 1975 *Federal Register* (pp. 28242–286), was developed to provide a public review period in which applicants could rebut EPA's finding against registration of a pesticide. In principle, the RPAR process was to last 300 days, and would encompass data analysis, public comment, risk-benefit analysis, USDA comment, and a final decision; in practice, all actual RPAR's took from one to five years (see the *Federal Register*, July 1975, for a detailed discussion of RPAR; see also May, 1984).

Economic Analysis of Pesticide Regulations

It is clear from EPA's mandate that pesticide regulations are intended to take into account both social benefits and costs of pesticides. Economic analysis of these benefits and costs, and of the design of policies to restrict or prohibit the use of a pesticide, falls within the domain of *welfare economics*. A basic concept used in welfare economics is that the change in an individual's economic well-being due to an event can be measured in terms of the individual's willingness to pay to obtain the event (if it is a "good") or to avoid it (if it is a "bad"). If all individuals in society are categorized as consumers and producers, then changes in social welfare can be analyzed in terms of markets. The welfare of consumers typically is measured by using the concept of consumer surplus. In terms of a market diagram, such as figure 2–1, consumer surplus is the area ABP_1 below the demand curve D_1 and above the price P_1 paid for a good, using the supply curve S_1. It can be shown (see Just, Hueth, and Schmitz, 1982) that a change in consumer surplus due to a change in the market equilibrium is an approximate measure of the change in consumer welfare or willingness to pay. Producer welfare is measured in terms of an analogous concept of producer surplus, which is defined as revenue minus variable costs of production for the good the producer is selling in the market. In terms of the market diagram, producer surplus is the area above the supply curve S_1 and below the market price P_1 (area BEP_1 in figure 2–1). A change in the market equilibrium causes a change in producer welfare equal to the change in producer surplus.

If there were no externalities associated with agricultural production of food, the market for an agricultural good would be represented as it is in figure 2–1. The initial competitive equilibrium is at point B, and it can be shown that this equilibrium generates the maximum social welfare, measured as the sum of consumer surplus (area ABP_1) plus producer surplus (area BEP_1), in the absence of market failure resulting from externalities. The introduction of a pesticide that increases agricultural productivity and thus reduces the marginal and average costs of production would shift the supply curve from S_1 to S_2, and result in a new market equilibrium at point C. As a result, there would be an overall gain in social welfare equal to area $BCDE$. Consumers would gain area P_1BCP_2 because of the reduction in the price of food; the change in producer welfare would be the area P_1BE minus P_2CD. It should be observed that *producers as a group could gain or lose from the introduction of the pesticide, depending on the relative elasticities of supply and demand.*[1] The more inelastic the demand, the more price falls and the more likely producers are to lose. If the market price were fixed, however, as would be the case if the government were to support the market price or if the good were being exported to a large world market, producers would necessarily gain welfare, but consumers would neither gain nor lose.

1. The elasticities of supply and demand are formally defined as the percentage change in the quantity relative to the corresponding percentage change in price. These elasticities are thus closely related to the slopes of the demand and supply curves.

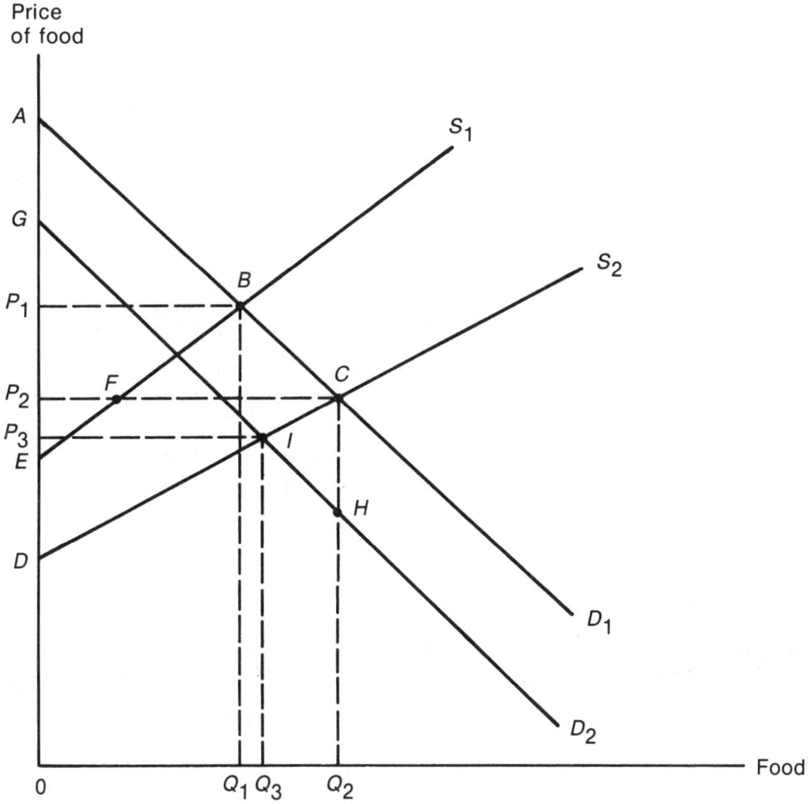

Figure 2–1. Welfare analysis in terms of consumer and producer surpluses

If the use of a pesticide resulted in externalities such as water pollution, consumer welfare would change in response to two effects. Welfare would be reduced because of the effects of the pollution, but would be increased by the reduction in the price of food that came from increased agricultural production. To represent the reduction in welfare due to pollution, the demand curve can be shifted downward from D_1 to D_2, as in figure 2–1. Thus demand curve D_2 reflects consumers' true willingness to pay, taking into account the pollution externalities resulting from pesticides used in food production. The socially optimal equilibrium in the market is now at point *I* rather than point *C*, where the equilibrium price and quantity of food are lower than at *C*. Because property rights typically are not assigned to common property resources such as water, the external costs of the pollution would not be accounted for in an unregulated free market. A free market, therefore, would equilibrate at *C*, not at *I*, and as a result social welfare would be lower than at *I*. At *C*, the total loss of consumer and producer surplus equals area *CHI*, reflecting the fact that by producing more than Q_3, resources are being used in a socially inefficient way because the marginal social cost of production, given by S_2, exceeds the marginal social benefit, given by D_2.

In cases where a free market fails to achieve the socially optimal resource allocation, a government agency such as EPA has two regulatory options: to modify the use of the pesticide through some policy instrument, or to ban its use altogether. If social welfare is lower at point *I* than at point *B*, the pesticide should be banned, since its use entails greater costs than benefits to society. But if social welfare is greater at *I* than at *B*, the regulating agency should devise policies to move the market equilibrium from *C* to *I*. For example, either consumption or production could be taxed so as to move the market equilibrium from Q_2 to Q_3.

The preceding analysis of consumer and producer surpluses is conducted in terms of market equilibrium. Market equilibrium analysis must be contrasted with analysis of the individual firm, which is the principal focus of this study. Ignoring the important differences between these two cases can lead to erroneous analysis and apparent paradoxes. For example, it is noted above that the introduction of a pesticide which reduces production cost does not necessarily increase producer welfare *in the aggregate*. It is logically possible that aggregate producer welfare could be reduced in some instances; since demand for food products generally is quite inelastic, this outcome is not unlikely. In such a case, the ban of a pesticide would increase aggregate producer welfare. Yet, agricultural interests generally oppose regulation of pesticides. This apparent paradox is resolved by recognizing that for the individual firm, at a given product price, the introduction of a pesticide that reduces average production cost or reduces production risk does increase the firm's welfare. The aggregate effect of this innovation, however, is to encourage production which leads to a lower market price and possibly lower producer welfare.

Another important dimension of the analysis of the effects of pesticides on producer welfare is recognition that the use and regulation of pesticides may have differential effects on producers who are geographically dispersed. Producers of a crop in region A, for example, may have a particular pest problem that producers in region B do not have. Producers in region A who have the pest problem may need to use a particular pesticide in order to compete with producers in region B; banning the use of that pesticide would harm producers in region A, but would not hurt producers in region B, and could even help them if the resulting reduction in production in region A caused the market price of the crop to increase. One important implication of this heterogeneity is that a uniform federal restriction of a pesticide is likely to be inefficient, because the optimal degree of regulation under one set of conditions generally will not be optimal under a different set of conditions.

It is worth noting, in this connection, that government policies in the United States and other countries often support the prices of major agricultural commodities, and that since many agricultural commodities are traded in world markets, the agricultural sector of even a large country faces a much more elastic demand curve than would be the case without trade. The more elastic the demand, the less a price falls as a result of the productivity effects of pesticides, and the more likely are farmers to gain from the use of pesticides. Thus, under conditions of a very elastic demand the restriction of pesticide use is likely to reduce producer welfare.

It is also relevant that in the analysis of pesticides the validity of the conventional measure of producer surplus can be questioned. Pesticides are different from other innovations in that they do not necessarily directly reduce the average cost of production on a given farm. Rather, they reduce the *likelihood* that the farmer's crop will be harmed by a pest. Under these conditions, it is not clear how the conventional measure of producer surplus, defined with respect to a deterministic output, is related to the change in welfare resulting from a reduction in production risk. Thus the apparent paradox described above may also be resolved if producer surplus is found not to be the appropriate measure of the value of pesticides to producers. Indeed, research has shown that conventional producer surplus is a valid measure of producer welfare under uncertainty only when very restrictive assumptions are made (see Pope, Chavas, and Just, 1983).

Neoclassical Analysis of Economic Efficiency and Producer Welfare

Conventional neoclassical production theory is the basis for the producer surplus concept introduced above. Neoclassical economic theory posits firms as profit maximizers within a world of certainty and costless information. Within this model, the firm's long-run decisions involve the choice of production technique and associated capital stocks. Given a production technique, a fixed capital investment, and all relevant technological and economic information, the only decisions remaining are the choice of optimal variable input and output levels.

The analysis of economic efficiency involves examination of the degree to which a firm maximizes the economic returns to its production activities. In the short run, the firm (given its technology, capital stocks, and fixed costs) is assumed to maximize economic profit, which is equivalent to maximizing producer surplus. Therefore, *analysis of economic efficiency is equivalent to the analysis of producer welfare in the short run*. The connection between efficiency and welfare of the firm is used throughout this study to develop methods for analyzing the efficiency and welfare effects of pesticide policy.

In neoclassical theory, the firm's production technology is described in terms of the production function which shows the maximum output obtainable from a given input set; thus the production function is said to indicate technical efficiency. A firm should choose the production technique (that is, the set of production practices and capital stocks) which is associated with the technically most efficient production function, because it gives the firm the most output and hence the most profit for a given cost. The contrast between an inefficient production function, f_0, and an efficient production function, f_1, is shown in figure 2–2. If a firm were choosing among pest management practices, it would choose those associated with the most efficient production function.

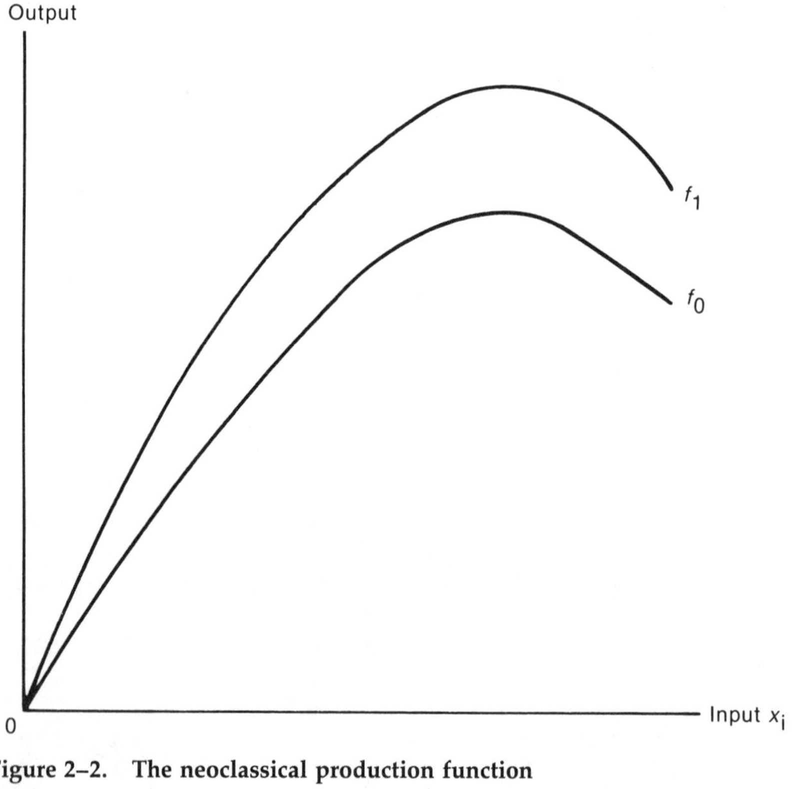

Figure 2–2. The neoclassical production function

In order to maximize profits, each input must be used up to the point that the additional ("marginal") benefit of using an additional unit of the input equals the additional (marginal) cost of the input. The marginal benefit (value of the marginal product) of an input equals the change in output resulting from an additional unit of input (marginal product of the input) times the output price. Letting p be output price and MP_1 be the marginal product of input x associated with production function f_1, the value of the marginal product is $VMP_1 = p \cdot MP_1$. The marginal cost of the input is simply the input price w when the firm purchases inputs in a competitive market. The graph of VMP_1 is shown in figure 2–3. There the VMP curve is decreasing to represent the typical relation between inputs and outputs known as the "law of diminishing returns": as more of input x is applied (holding other inputs constant), beyond some point the additional amount of output gained from a unit of input declines. The economically efficient (profit maximizing) quantity of input is given by the intersection of the VMP curve and the $w = w_0$ line. Note that

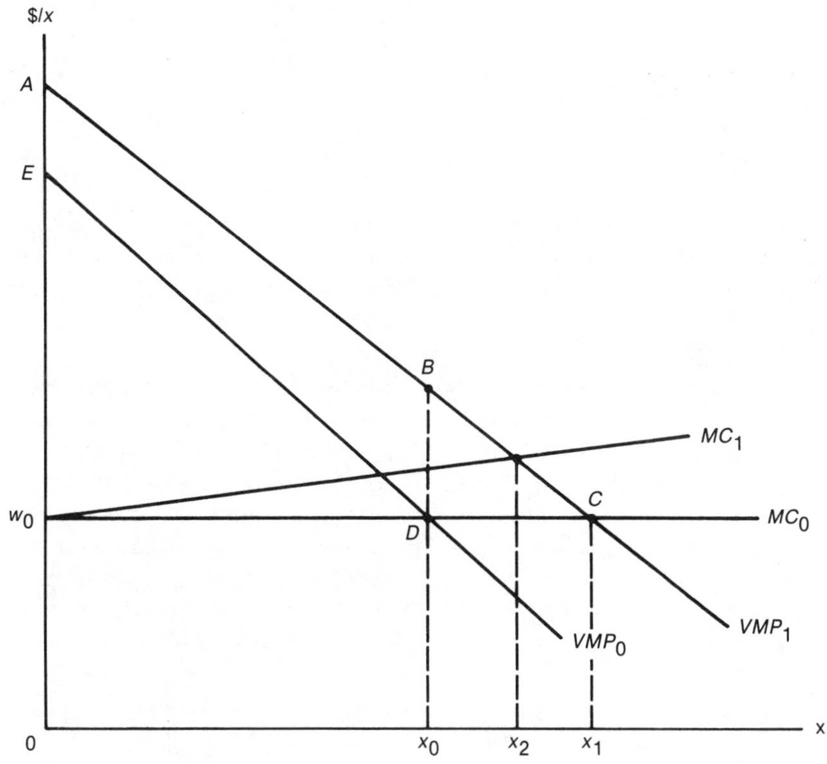

Figure 2–3. The VMP curve and producer surplus in the neoclassical model

the firm can buy as much of x as it wants to at price w, so the horizontal line at w_0 is the marginal cost of the input or the supply curve of x to the firm. The firm that uses inputs to equate the factor price to its *MVP* is said to be allocatively efficient. Thus the allocatively efficient firm maximizes profit, given its production function and the prices it faces.

The net benefits to the firm of using an input can be measured as the revenue the input generates minus the input cost. Since the *VMP* curve measures the additional revenue that each additional unit of input generates for the firm, the area under the *VMP* curve represents the total revenue that an input generates. The cost of that input equals the price times quantity, or the area under the factor supply curve. Thus the net benefits to the firm of using an input equals the area below the *VMP* curve and above the factor supply curve. This area is equivalent to the measure of producer surplus, defined as revenue minus variable factor cost.

The welfare loss to the firm resulting from technical or allocative inefficiency can be measured in terms of producer surplus, as illustrated in figure 2–3. If the firm is technically and allocatively efficient, it produces with the efficient production function f_1 of figure 2–2, and thus uses the associated schedule VMP_1 to choose its optimal input level. If the firm is technically inefficient and produces with production function f_0 of figure 2–2, its *VMP* curve would be VMP_0 in figure 2–3, and its efficient input use would be x_0. The loss to the firm from technical inefficiency is equal to the loss in producer surplus—that is, the area between the two *VMP* curves and above the factor supply curve (area *ACDE*). If the firm is technically efficient but allocatively inefficient and uses x_0 instead of x_1, the loss in producer surplus equals the triangular area between VMP_1 and the factor supply curve to the right of x_0 (area *BCD*).

The concepts of economic, technical, and allocative efficiency are useful in the analysis of pesticides and pest management. Allocative efficiency can be used to evaluate the welfare effects of a government regulation restricting input use. If the firm is allocatively efficient, and a pesticide use restriction is binding, it would be forced to use less of the pesticide and thus face a loss in welfare equal to the corresponding decline in producer surplus. However, if the firm is not initially allocatively efficient, the regulation could either increase or decrease the firm's welfare, depending on where the firm is producing in relation to the efficient point.

The concept of technical efficiency can be used to analyze the effects of the introduction of a new production technique on producer welfare. For example, the introduction of an IPM technology which increases a farmer's ability to manage pests would result in an increase in technical efficiency, and thus an increase in welfare equal to the resulting gain in producer surplus.

Several additional observations are in order about the usefulness of the neoclassical theory of production for the analysis of pest management. First, if the relevant benefits and costs are defined in terms of a firm's revenues and costs, the neoclassical solution to the pesticide use problem is the same as the economic threshold solution: pesticides are used to the extent that the marginal benefit exceeds the marginal cost. However, if the economic threshold were replaced by a more broadly defined ecological threshold that took into account the ecological externalities associated with the pesticide, the amount of pesticide used would generally be less than the amount prescribed by the neoclassical economic model. For example, if the ecological threshold defined costs to include the future effects of a pesticide on pest resistance, and if these effects were not incorporated into a farmer's private production decisions, the marginal social input cost might resemble, in figure 2–3, MC_1 rather than MC_0, indicating that the marginal resistance cost was increasing in the amount used. Therefore, the ecological threshold would indicate that only quantity x_2 should be used, not x_1.

Second, in the static, certain world of neoclassical theory there is no time dimension in the analysis of optimal input use. Thus the neoclassical theory cannot be used to analyze pesticide use within the dynamic context of the physical and biological environment emphasized by Flint and van den Bosch in their description of integrated pest management. Within the static neoclassical model, the decision maker has perfect and complete information, so there is no need for the information gathering, assessment, and dynamic decision rules for pesticide application that Flint and van den Bosch view as essential elements of an IPM technology.

And third, the neoclassical theory assumes away the uncertainty of the real world. Outside of neoclassical theory, uncertainty is recognized as a major problem faced by farmers, and pests represent one of the major elements of farm production uncertainty, along with weather. Indeed, it could be argued that the uncertainty associated with agricultural pests is the principal consideration in farmers' choices of pest management practices and in their pesticide use decisions. The neoclassical theory of production therefore may be inappropriate to evaluate the efficiency of alternative pest management practices or of farmers' pesticide decisions.

Producer Efficiency and Welfare Under Uncertainty

In the neoclassical theories of consumer and producer behavior, the welfare analysis of the consumer and the firm are different in a very important respect: the consumer makes choices to maximize utility which is not directly observable, while the firm chooses inputs to maximize profit which is observable. This difference means that the measurement of producer welfare in terms of a money value is direct, whereas the measurement of consumer welfare cannot be direct, and must be done indirectly either in relation to measured demand functions or elicited consumer preferences.

When a firm's prices and output are uncertain at the time the firm makes its decisions, the firm cannot be assumed to be maximizing profit because profit is not a function with a well-defined maximum. Therefore, under uncertainty, the firm must be assumed to be maximizing some function which relates its input decisions to its welfare. In the theory of the firm under uncertainty, this is done by assuming that the firm maximizes an objective function involving the relation of its input decisions to the probabilities of the various profit levels that are possible ex ante. The most common objective function in such research is the maximization of the mathematical expectation of the utility of profit or wealth. When this assumption is made, the analysis of producer welfare under uncertainty is similar to the neoclassical analysis of the consumer, and similar theoretical difficulties arise.

Under uncertainty there are two approaches to the analysis and measurement of producer welfare. The more conventional approach is to estimate the appropriate output supply or factor demand functions, taking into account the effects of uncertainty on them, and then using these estimates to calculate producer surplus or some analogous money-equivalent measure of welfare in terms of areas above or below the functions. The alternative approach is to attempt to directly measure the parameters of the decision maker's objective function. The estimated objective function is then used to compute welfare changes in terms of utility or its money equivalent.

In view of the widespread use of the expected utility theory to analyze decision making under uncertainty and to analyze and measure producer welfare, it is useful to briefly digress on the elements of this theory. Many of the associated concepts of utility theory are central to the subject of this study.

Expected Utility Theory and Risk Attitudes

Much has been written about the meaning of risk and uncertainty. Within the modern analytical framework that is used to model decision making under uncertainty, defining risk and uncertainty is straightforward. For the purposes of this study, risk and uncertainty can be thought of as equivalent concepts, and mean that some variables that affect the welfare of an individual decision maker are random variables. To illustrate the meaning of a random variable, consider a simple stylized model of pest management. The farm manager makes a decision at time 1, such as the choice of a pest management action, which is intended to affect crop yield at time 3. This decision influences production but does not entirely determine it, because other events which cannot be accurately predicted or controlled by the manager—such as the effect of insect pests—occur at time 2 and also affect crop yield. Consequently, it is said that the farm manager faces a probability distribution of yields associated with each decision, rather than a certain yield. This distribution indicates the probabilities associated with each possible yield outcome. Three such hypothetical yield distributions are presented in table 2–1, which shows the probability of each yield falling within each of five intervals. Where a manager chooses to take no action, for example, the yield will fall between 0 and 100 with a 1 in 10 chance, or with a 0.1 probability. The choice of a scheduled spray program decreases the probability of a low yield and increases the probability of a higher yield; that is, the choice of spraying skews the result toward higher yields. Thus, when the farm manager makes decisions under uncertainty, he is not choosing one certain outcome, but rather is choosing a set of possible outcomes that occur with some probability. These sets of outcomes, or probability distributions, define a random variable such as yield. Management decisions under uncertainty therefore involve choices between alternative probability distributions of uncertain outcomes.

In order to choose among distributions of outcomes, the decision maker must know what those distributions are. But often the "true" or "objective" distributions may not be known by the decision maker. In his famous book *Risk, Uncertainty, and Profit*, Frank Knight defines risk in terms of random events with known probabilities, and uncertainty as events with unknown probabilities. Modern decision theory makes this distinction unnecessary by assuming that individuals have subjective beliefs about the distributions they are choosing. The "subjective" distributions need not correspond to the objective ones. It is often argued that decision makers learn over time about the objective distributions which generate observed phenomena, and that they

Table 2–1. Hypothetical Yield Distributions and Pest Management Decisions

	Probabilities of net returns between				
Management decision	0–100	100–200	200–300	300–400	400–500
No action	.10	.40	.30	.17	.03
Follow spray schedule	.05	.35	.30	.25	.05
Use IPM program	.05	.30	.40	.20	.05

update their subjective beliefs over time according to their observations. While it seems unlikely that decision makers always know the objective distributions, it seems equally unlikely that those having economic incentives to learn them will not make efforts to do so. Decision makers are said to form rational expectations when their subjective distributions are assumed to be equal to the objective distributions generating observed phenomena. In this study rational expectations are assumed.

The fundamental problem facing the manager, then, is to choose among the alternative random variables (probability distributions) associated with each management option. To analyze how such decisions are made, a variety of models of decision making under uncertainty has been devised (for a survey of such models, see Anderson, 1979). The principal model used by economists, and the one discussed here, is based on the work of von Neuman and Morgenstern (1947). Underlying the von Neuman-Morgenstern approach is the utility function—a schedule showing how the decision maker ranks all of the possible outcomes. The value of each outcome is referred to as its utility; two representative utility functions are shown in figure 2–4. Von Neuman and Morgenstern hypothesize that under uncertainty

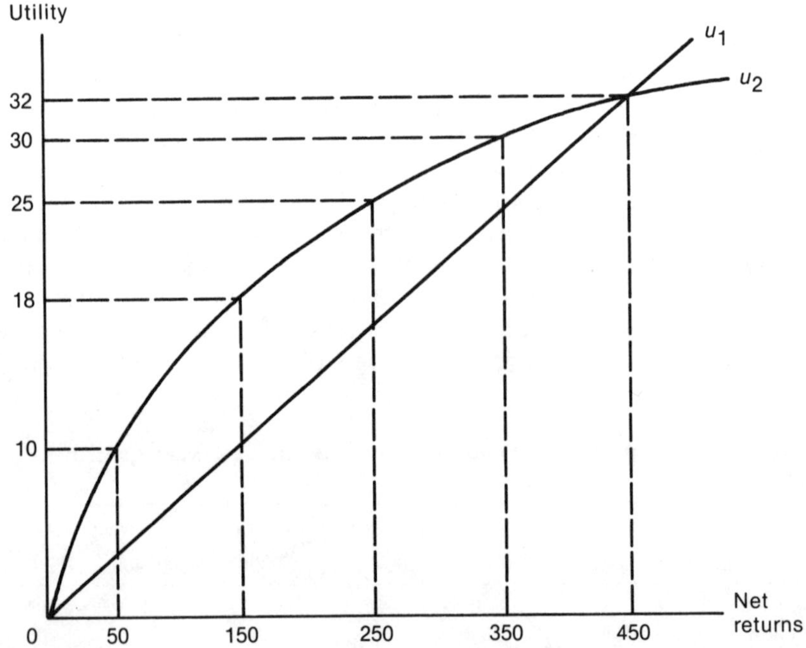

Figure 2–4. The utility function of a risk-averse decision maker

rational individuals make choices based on the utility they associate with each outcome and its probability. Specifically, von Neuman and Morgenstern show that given a set of simple axioms, rational individuals order probability distributions by comparing the probability-weighted average of the utility of all possible outcomes. The probability-weighted average of utilities is known as an expected utility.

Economists use the expected utility principle to describe the risk attitudes of decision makers. A risk-averse individual obtains greater expected utility from a sure outcome than from a risky outcome with an average value equal to the sure outcome; a risk-loving individual obtains less expected utility from a sure outcome than from a risky outcome; and a risk-neutral individual obtains the same expected utility from a sure outcome as from a risky outcome. These definitions of risk attitudes can be shown to correspond to the value an individual attaches to insurance. A risk averter is willing to pay a positive amount to insure against risk; a risk lover is willing to pay to take a risk, and would not buy insurance. The maximum amount that an individual would be willing to pay for insurance is defined as the risk premium. That premium is positive for a risk averter, zero for the risk neutral, and negative for a risk lover.

It can also be shown that a risk-neutral decision maker's utility for each outcome is essentially the dollar value of that outcome. In figure 2–4, a risk-neutral utility function u_1 is illustrated by a straight line; such a linear utility function means that a gain or loss of income is valued the same to the individual, regardless at what income level that gain or loss occurs. That individual's utility of an outcome, then, is equivalent to the number of dollars associated with the outcome, since utility is a linear transformation of the dollar value of the outcome. Thus, according to the expected utility principle, the risk-neutral manager takes that decision associated with the greatest expected utility, which in turn equals the outcome with the greatest average value.

The risk-neutral case can be illustrated by turning to table 2–1 and calculating the mathematical expectation of each yield distribution, as follows. Where the decision is for no action, the expected net return (calculated at the midpoint of each interval) is $50(.1) + 150(.4) + 250(.3) + 350(.17) + 450(.03) = 213$; by similar calculations, the mean for scheduled spraying is 240, and for the IPM program is also 240. In this example, the risk-neutral manager would prefer to use either scheduled spraying or the IPM program rather than take no action, but would be indifferent to the choice between scheduled spraying and the IPM program.

A risk-averse decision maker systematically values outcomes differently than a risk-neutral decision maker does. Specifically, the risk averter values increments to an outcome at a decreasing rate. An example of such a utility function is shown in figure 2–4 as u_2. The utility function of a risk averter is concave, indicating that the value of an additional dollar of return is worth less at high income levels than at low income levels. This type of behavior is the basis of the risk averter's willingness to buy insurance, since the chance of an unusually high gain is worth less than the cost of a large loss. The risk averter is thus willing to trade off some income (the price of insurance) to guarantee against the chance of large losses. Applying the expected utility approach with the utility function u_2 in figure 2–4 to the decision problem in table 2–1, the expected utility of taking no action is $10(.1) + 18(.4) + 25(.3) + 30(.17) + 32(.03) = 21.76$, but the expected utility of scheduled spraying is 23.4, and of the IPM program, 23.6. This expected utility analysis predicts that the risk-averse manager would choose the IPM program. It should be noted that the units of utility are arbitrary, and generally do not correspond to monetary values.

Production Uncertainty, Input Use, and Producer Welfare

The concepts of production risk, defined in terms of randomness in production, and the farmer's risk attitudes, defined in terms of the utility function, can be used to generalize the neoclassical efficiency and welfare analyses. The distribution of output, conditional on management decisions, replaces the neoclassical production function. The maximization of expected utility of wealth or profit replaces the profit maximization postulate of the neoclassical model. Expected utility is maximized by choosing the input level at which an additional (marginal) unit of input gives no higher utility—that is, the input level at which expected marginal utility (EMU) equals zero. Figure 2–5 shows several EMU functions. These functions slope downward under the assumption that the expected utility function is globally concave in inputs. For the manager having EMU_0, the optimal input level would be x_0. Thus the area under the EMU curve measures the firm's welfare in terms of utility. A deviation from x_0 in input use would result in a loss in welfare equal to the loss of utility. It follows that using x_2 instead of x_0 would result in a welfare loss measured as the area BCx_0x_2.

It is also possible to view the model based on expected utility maximization as a generalization of the neoclassical model. Figure 2–6 shows the factor supply curve to the firm and the expected value marginal product curve (EVMP) for the input. The risk-neutral firm will use the input up to the point that the expected value marginal product equals the factor price—x_0 in figure 2–6. The area above the

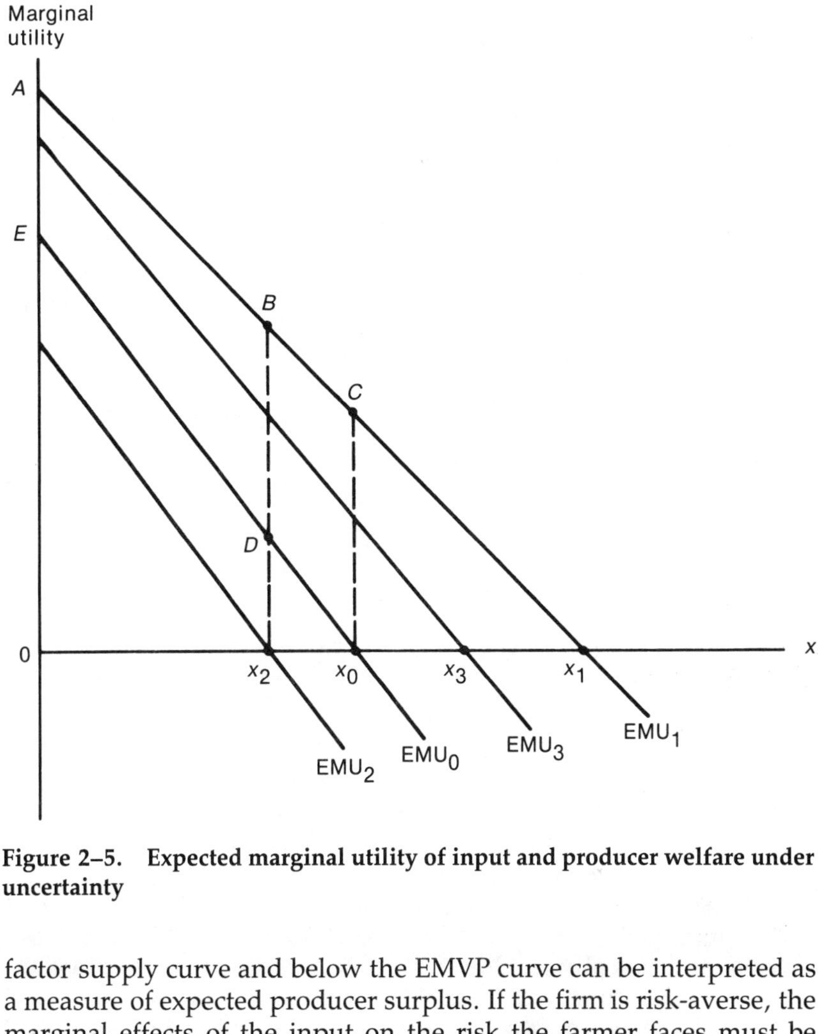

Figure 2–5. Expected marginal utility of input and producer welfare under uncertainty

factor supply curve and below the EMVP curve can be interpreted as a measure of expected producer surplus. If the firm is risk-averse, the marginal effects of the input on the risk the farmer faces must be taken into account. Following Pope and Kramer (1979), the input can be defined as marginally risk-increasing (reducing) if the risk-averse firm uses less (more) of it than the risk-neutral firm. Therefore if the input is risk-increasing, the marginal benefit curve of the input will lie to the left of EVMP, as does the curve labeled $SVMP_1$, and the firm will use quantity x_1 of the input; if the input is risk-reducing, the curve will lie to the right of EVMP, as does $SVMP_2$, and optimal input use will be x_2. If the SVMP curves are appropriately adjusted to take into account the effects of risk on expected utility, the firm's welfare can be measured as the area above the factor supply curve and below the SVMP curve. Strictly speaking, the conventional factor

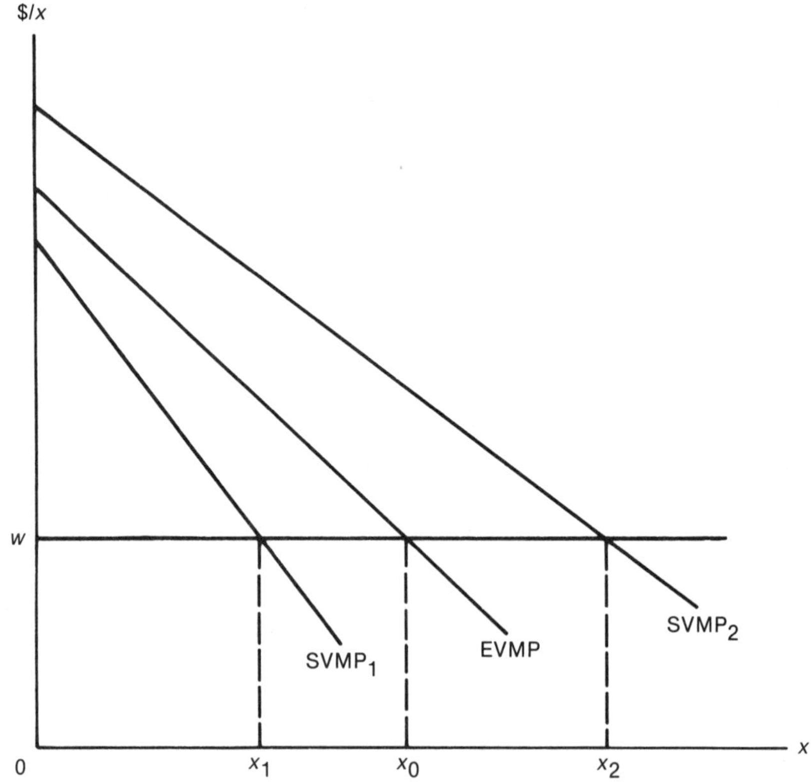

Figure 2–6. Producer welfare under uncertainty and the EVMP curve

demand curve under risk is appropriate only if the decision maker's preferences exhibit constant absolute risk aversion, as shown by Pope, Chavas, and Just (1983). Otherwise, the demand curve must be adjusted to reflect the effect of wealth changes on welfare.

It should be noted that the vertical distance between the EVMP curve and the SVMP curve indicates the contribution of risk to the marginal value of the input to the firm. In particular, in the case of $SMVP_1$ it can be seen that risk effectively causes the marginal value of the input to be less than if the farmer were risk neutral. In other words, it would require a subsidy equal to the vertical distance between $SVMP_1$ and EVMP to induce the risk-averse farmer to behave as if he were risk neutral. This vertical distance can be defined as the marginal risk premium associated with the input. If the input is marginally risk-increasing, the marginal risk premium is positive; if the input is marginally risk-decreasing, the marginal risk premium is negative.

It also should be noted that figures 2–5 and 2–6 tell the same story in terms of input demand behavior and firm welfare. The point at which the EMU curve crosses the horizontal axis in figure 2–5 corresponds to the point at which the SVMP curve crosses the factor supply curve in figure 2–6. However, in figure 2–5 the area under the EMU curve represents firm welfare measured in units of utility, whereas the area measured in figure 2–6 represents money-equivalent units. That is to say, if there were a change in the firm's welfare as measured in utility according to figure 2–5, the firm would value that change in dollar terms according to the areas below the factor demand curve in figure 2–6.

Using this framework for welfare analysis, the firm's allocative and technical efficiency can be examined in a manner analogous to the analysis of efficiency in the neoclassical model. In the expected utility model, allocative efficiency can be evaluated by comparing, in figure 2–5, the area under the EMU curve from zero to the optimal input level (where the EMU curve cuts the horizontal axis) with the area under the EMU curve from zero to the actual input level. If x_0 is used when x_1 is optimal, for example, the allocative inefficiency results in a welfare loss to the firm that is measured as area Cx_1x_0. Technical efficiency under uncertainty is defined in terms of the efficient output distribution. If the firm produces with the efficient distribution, it will attain the greatest utility possible from its inputs. Thus if efficient production gave EMU_1 and inefficient production gave EMU_0, the welfare loss due to technical inefficiency would be area Ax_1x_0C. If the firm produces with an inefficient output distribution, its utility is lower than with the efficient distribution and the area under the inefficient EMU curve is less than the area under the efficient one. The most efficient distribution need not be associated with a higher input level, hence the EMU curves can cross.

A fundamental difference between neoclassical efficiency analysis and the analysis of efficiency under uncertainty is that in the former case the welfare loss from inefficiency depends on the firm's technology and prices. However, *under uncertainty, efficiency also depends on the firm manager's risk attitudes, as represented by the utility function.* This is demonstrated by the fact that the position of the EMU curves in figure 2–5 depends on the effect the input has on the output distribution (the technology) and on the decision maker's risk attitudes. It follows that even if two farmers are producing with the same input quantities and the same technology, their efficiencies can be different if their risk attitudes differ. Suppose, for example, that two farmers have different utility functions but use the same technology, and both produce with input x_0. If x_0 were efficient for one of the farmers, it could not be efficient for the other because he has a different utility function. Therefore *both the risk attributes of the technology and risk attitudes of the farmer must be known in order to assess efficiency and welfare under uncertainty.*

Production Dynamics, Sequential Decision Making, and IPM

The neoclassical model also needs to be generalized to account for the dynamic dimension of agricultural production and decision making. The definition of IPM and the preceding discussion of decision making under uncertainty suggest that the time dimension is an important element in the pest management process. Following that line of reasoning, the pest management decision-making process can be represented schematically, as in figure 2–7. As the production process proceeds over time, input decisions x_t (that is, pest management actions) must be taken (including the decision to take no action). Also over time, pest events ϵ_t occur that are random from the manager's point of view. At each decision point the manager uses information Ω_t. The essential question is how the manager makes these decisions over time, given the information available at each point in time and the likelihood of future random events.

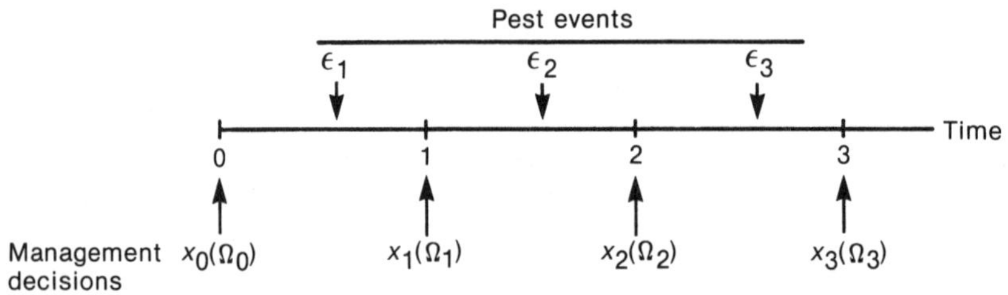

Figure 2–7. Analysis of sequential decision making

Consider first the case of a pest management program involving a predetermined schedule of actions (applications of pesticides, cultural practices, and the like). Since this schedule is predetermined, it can be said to be based on a priori information available to the manager before production begins, and thus is represented as Ω_0. In a scheduled program, then, at each time t action $x_t(\Omega_0)$ is taken, regardless of the pest events ϵ_t that occur over time.

In contrast, an integrated pest management program would involve the continuous accumulation of information about the status of the crop and pest populations, and pest management actions based on that information. Thus in an IPM program action $x_t(\Omega_t)$ is taken at time t, where $\Omega_t = (\Omega_{t-1}, x_{t-1}, \epsilon_t) = (\Omega_0, x_0, x_1, \ldots, x_{t-1}, \epsilon_1, \epsilon_2, \ldots, \epsilon_t)$. In words, in an IPM program each action is based on all previously available information, and on the histories of pest events and pest management actions.

In terms of efficient input utilization, the EMU curves are now functions of the relevant information used in making input decisions. Figure 2–8 illustrates the differences between a scheduled pest management program and an IPM program. The scheduled program is based on the information set Ω_0, and input decisions are made according to the expected marginal utility curve $EMU_t(\Omega_0)$. In contrast, when input decisions are made sequentially in an IPM program, they are based on curve $EMU_t(\Omega_t)$, which may lie to the right or to the left of $EMU_t(\Omega_0)$, depending on the pest events and pest management actions that are optimal.

This analysis has two important implications for the understanding of pest management efficiency. First, given the information that is

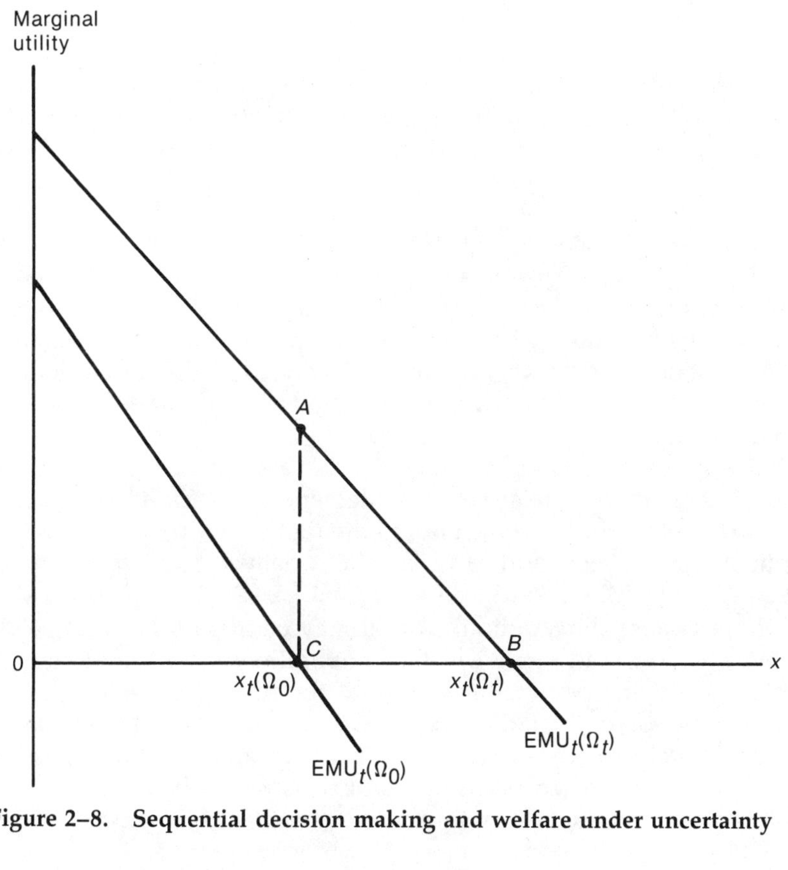

Figure 2–8. Sequential decision making and welfare under uncertainty

available over time, the scheduled approach to pest management can never be more efficient than the IPM program based on information updating, and generally will result in less efficient input utilization, as indicated in figure 2–8 by the divergence between $x_t(\Omega_0)$ and $x_t(\Omega_t)$ and the associated welfare loss *ABC*. Second, the IPM program does not necessarily result in the use of less pesticide by an individual farm manager than would be used with a predetermined pesticide application schedule. It is logically possible that in some cases more pesticides would be applied with an IPM program. This could occur, for example, if there were a greater than expected pest infestation, such that $x_t(\Omega_t) > x_t(\Omega_0)$. However, if farmers are risk averse, it is likely that they will prefer to err on the side of caution, so that in the aggregate more pesticides would be used in a scheduled spray program than in an IPM program.

Efficiency Concepts Compared

The efficiency concepts discussed above have been defined in terms of the benefits and costs to the firm using the pesticide. Such benefits and costs are referred to as "private." Efficiency from society's point of view must take into account these private effects of pesticide use as well as the effects on all other members of society. If all of the additional "social" benefits and costs are considered, the socially efficient level of pesticide use—where marginal social benefits equal marginal social costs—could differ from the firm's profit or utility maximizing levels. To illustrate, consider again figure 2–5, where EMU_0 represents the risk-neutral farmer's input demand curve, based only on the benefits to the farmer using the input. If the farmer had to pay not only the market price of the pesticide but also the additional social costs, the EMU curve might be shifted to EMU_2, and the socially optimal input level would be x_2, not x_0.

But suppose the farmer actually is risk averse and the input is risk-reducing, so that the relevant factor demand curve from the farmer's point of view is EMU_1, which lies to the right of EMU_0. The farmer's optimal input level would be x_1, but the socially optimal input would be x_3.

These relationships indicate why it may be important to take risk into account in analyzing the welfare implications of policies designed to bring farmers' pesticide use close to the socially optimal level. If risk were ignored, the policy analyst would conclude that the socially optimal level of pesticide use is x_2. If a farmer were risk averse, a policy which restricted pesticide input to x_2 would be inefficient for both society and the farmer, and could be worse from society's point of view than no regulation. Moreover, in this case the policy analyst would underestimate the welfare costs to the farmer of restrictions on pesticide use. In figure 2–5, if EMU_0 was mistakenly believed to be relevant to the farmer, and input use was restricted to x_2, the policy analyst would conclude that the welfare loss to the farmer because of the restriction was area Dx_0x_2. But if the correct demand curve was EMU_1, the true cost to the farmer of the restriction would be $Bx_1x_2 > Dx_0x_2$. Conversely, if the policy analyst assumed that the farmer was risk averse but the farmer actually was risk neutral, the policy would incorrectly restrict pesticide use to x_3 rather than to the socially optimal level x_2, which would have no effect on the farmer's welfare but would allow pesticides to be overused from society's point of view.

The conclusion to be drawn from this analysis is that in order to make an accurate evaluation of the private and social costs of pesticide use, to understand the technical and allocative efficiency of agricultural producers, and to estimate the welfare effects of pesticide policies, it is necessary to measure both the stochastic production technology and producers' risk attitudes.

Notes on the Literature

A standard reference to neoclassical production theory is Ferguson (1971). The literature on pest management is vast; Flint and van den Bosch (1981) is a good starting point. For bibliographies of economic analysis of pest management, see Osteen, Bradley, and Moffitt (1981) and McCarl (1981). Useful references for modern welfare economics are Just, Hueth, and Schmitz (1982) and McKenzie (1983). For an analysis of pest management issues in a welfare economics framework, see Council for Agricultural Science and Technology (1980). Discussions of and references to the literature on decision making under uncertainty can be found in Anderson, Dillon, and Hardaker (1977), Diamond and Rothschild (1978), and Lippman and McCall (1981). Newbery and Stiglitz (1981) provide an excellent discussion of the firm under uncertainty. For an overview of dynamic optimization models and sequential decision making, see Kendrick (1981).

three

THEORETICAL FOUNDATIONS

This chapter discusses in greater detail the theoretical issues related to the analysis of pest management technology that were raised in the previous chapter. A brief discussion of pest management as interpreted in the neoclassical model of production is followed by generalizations of the neoclassical theory. Topics discussed include the modeling of stochastic production, production dynamics and sequential decision making, the measurement of efficiency and producer welfare in the stochastic case, and the measurement of technological change under uncertainty.

Pest Management Technology in the Neoclassical Model

The neoclassical theory of production provides economists with their standard set of tools for analysis of production relations, and has been used by economists to measure and analyze pesticide productivity (see, for example, Headley, 1968; Carlson, 1977). As the neoclassical framework is familiar to most economists and provides insight into the issues involved in the analysis of pest management productivity, it is useful to begin this discussion within that framework. The particular limitations of the neoclassical model for analysis of pest management technology indicate the directions in which the neoclassical model needs to be generalized.

In the neoclassical model the competitive firm's short-run objective is specified as the maximization of profit, subject to the firm's production function and the fixed factors of production. The production function defines the maximum amount of output obtainable from the firm's technology and fixed inputs, and is written

$$q = f(x, z, \phi), \tag{3.1}$$

where q is output, x denotes a vector of variable inputs (such as labor, pesticides, fuels, and water), z is the vector of fixed inputs (such as land and other physical capital, and the farm manager's human capital), and ϕ is a vector of parameters of the production function. The function f is assumed to satisfy certain properties, such as monotonicity, quasi concavity, and continuity in x. The firm's variable input choice problem is defined as

$$\max_x J(x, z, \alpha) \text{ subject to } J(x, z, \alpha) = pq - wx, \quad q = f(x, z, \phi), \tag{3.2}$$

where p is the output price, w is a vector of input prices conformable to x, and $\alpha = (p, w, \phi)$ (the usefulness of this notation will be apparent in the generalization of the neoclassical model discussed below). $J(x, z, \alpha)$ is defined as the returns to the fixed factors of production, and thus is equal to the firm's producer surplus as defined in the previous chapter. For some purposes (for example, analysis of shutdown decisions) it is important to distinguish between profit (revenue minus total cost) and producer surplus (revenue minus variable cost). For the purposes of this discussion, however, this distinction is not important (see Just, Hueth, and Schmitz, 1982, chap. 4).

For interior solutions to equation (3.2), the first-order condition is

$$p\frac{\partial f(x, z, \phi)}{\partial x_k} = w_k, \quad k = 1, \ldots, n \tag{3.3}$$

Condition (3.3) can be resolved for the factor demand functions if the conditions of the implicit function theorem are satisifed. In this case there exists a system of equations

$$x_k = x_k^*(p, w, z, \phi), \quad k = 1, \ldots, n \tag{3.4}$$

which satisfy (3.3).

Equation (3.3) states the optimality condition for variable input use in the neoclassical model: the firm uses each input in such a way that the value of its marginal product, given by the left-hand side of (3.3), equals the factor price. When the firm uses the input quantity defined by (3.3) it is said to be *allocatively efficient*. If, given its input choices, the firm obtains as much output as is feasible with its production technology, it is said to be *technically efficient*. A firm that is both allocatively and technically efficient is *economically efficient*. The meaning of economic efficiency in the neoclassical model is apparent from

(3.2): the economically efficient firm obtains the maximum possible economic returns to its fixed inputs.

Efficiency can be measured in various ways. The conventional approach to efficiency analysis was originated by Farrell (1957), and has been generalized and refined in the recent literature (see Forsund, Lovell, and Schmidt, 1980; Kopp, 1981). The Farrell analysis of efficiency is conducted in terms of the effects that technical and allocative efficiency have on the firm's cost of producing a given output rate. Textbook economic theory emphasizes that the cost minimization problem at a given output rate is formally equivalent to the output maximization problem at a given total cost. This fact suggests that efficiency can be measured in terms of output rather than total cost, as has been done by Timmer (1971) and by Herdt and Mandac (1981). More generally, the profit maximization assumption of neoclassical theory suggests that efficiency analysis can be based on the firm's profit. As noted above, in the short-run case efficiency analysis based on profit is equivalent to defining efficiency in terms of quasi-rent or producer's surplus. We turn now to the analysis based on profit as it leads naturally to the more general concepts needed for the analysis of efficiency under uncertainty.

The use of a firm's short-run profit to measure allocative and technical efficiency is illustrated in figure 3–1. The efficient production function is $q^*(x, z, \phi)$. Under neoclassical assumptions, q^* is concave in inputs, so the corresponding objective function J^* is concave in inputs and has a unique interior global maximum defined by the conditions in equation (3.3). The technically and allocatively efficient firm therefore is at point A in figure 3–1.

Two interpretations can be given to technical inefficiency. The first is based on defining technical efficiency in terms of the frontier production function $q^*(x, z, \phi)$. The technically inefficient firm produces with the same production function q^* as the technically efficient firm, but for some reason the former obtains an output less than q^*, and thus earns less profit than the latter. As shown in figure 3–1, at input x_k^1 the technically inefficient firm realizes a profit associated with a point such as B′, in contrast to the technically efficient firm's profit at point B.

The second interpretation is based on the view that an inefficient firm produces with a production function which lies below the efficient one in input-output space. That is, the inefficient firm produces with a production function $q^i(x, z, \phi)$ such that $q^i(x, z, \phi) \leq q^*(x, z, \phi)$ for all x, with strict inequality for at least one x. This interpretation of technical efficiency underlies a large number of production function studies which have attempted to explain productivity differences in terms of conventional inputs as well as other variables, such as human

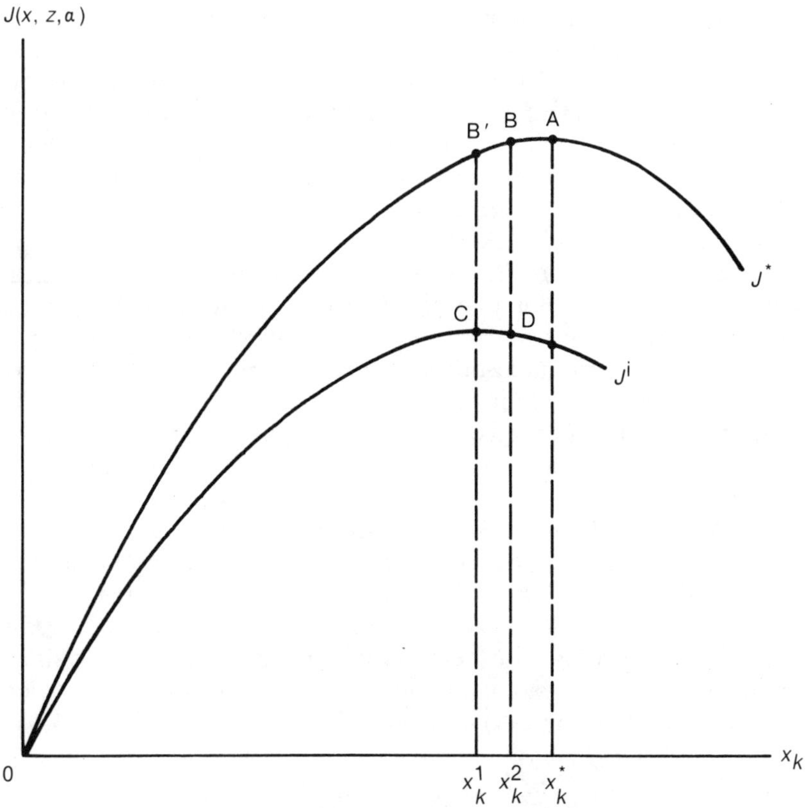

Figure 3–1. Efficiency analysis using a firm's objective function

capital (see, for example, Lockheed, Jamison, and Lau, 1980).

Allocative efficiency can also be measured in terms of the objective function. Given that production is efficient, in figure 3–1 the allocatively efficient input is x_k^*. Any other input level reduces the value of the firm's objective function—that is, it would cause a welfare loss to the firm.

These two interpretations of technical inefficiency have implications for the measurement of allocative efficiency. If a technology is defined as the frontier function q^*, then allocative efficiency can be measured only relative to the corresponding objective function J^*. But if a technology is defined as a set of more and less efficient production functions, allocative efficiency can be defined relative to each production function. If a firm produces with q^i, it chooses x to maximize $J^i(x, z, \alpha) = pq^i(x, z, \phi) - wx$. The allocatively efficient choice satisfies $\partial J^i(x, z, \alpha)/\partial x = 0$, not $\partial J^*(x, z, \alpha)/\partial x = 0$. Thus point C in figure 3–1 represents allocative efficiency given that the production function is q^i.

The analysis of efficiency using the firm's objective function can be related to the conventional Farrell approach. As Kopp (1981) shows, the Farrell measure of allocative efficiency can be interpreted as measuring a move along a given isoquant from the cost-minimizing factor proportions to the actual factor proportions. Total cost at the efficient point relative to total cost at the inefficient point is a measure of allocative efficiency. In figure 3–1, the move from the efficient to the inefficient point is equivalent to sliding down the objective function from point A to point B, since an increase in total cost implies a corresponding reduction in profit. However, in the objective function approach the move from the efficient point to the inefficient point does not necessarily hold output constant. Thus to make the Farrell efficiency measure and the objective function measure equivalent, the latter must be compensated for the output effect.

Interpreting technical and allocative efficiency in terms of the firm's objective function shows the direct connection between efficiency analysis and welfare economics. Given the firm's technology and inputs, and thus given the cost of production, the more output it can produce the better off the firm will be from the standpoint of producer surplus. Similarly, allocative inefficiency can be interpreted as reducing producer surplus. The connection between efficiency analysis and producer welfare is exploited later in this chapter to generalize efficiency analysis to the case of production uncertainty.

The Structure of Pest Management Technology

In the general neoclassical production function (3.1), all inputs generally are nonseparable in the sense that the marginal rate of technical substitution between any two inputs depends on the values of other inputs. But in many production processes the inputs are service flows from subprocesses; the inputs into the subprocess are separable from other inputs in the sense that the marginal rate of technical substitution between inputs in the subprocess is invariant to inputs not used in the same subprocess. Berndt and Wood (1979) have suggested that capital and energy often are combined in industrial subprocesses to yield capital services flows. In agriculture, various production operations such as field preparation, cultivation, and pest management can be thought of as distinct subprocesses.

As an application of the concept of separability of production subprocesses, suppose that pest management is a distinct activity with identifiable inputs and outputs. Define the pest management subfunction as

$$x_m = m(v_m, z_m), \tag{3.5}$$

where x_m is an element of x measuring the flow of pest management services, v_m is a vector of variable inputs into the production of pest management services (such as chemicals and human labor), and z_m is a vector of fixed capital inputs for pest management (such as machinery for insecticide sprays). This subfunction can be thought of as a production function with the same properties as the function (3.1). Pest management productivity can be defined as a measure of the efficiency of this subprocess in producing the service flow x_m. Since x_m is in turn an input into the crop production function (3.1), it is clear that pest management productivity affects overall crop productivity.

Separability in the production process is known to have important implications for producer behavior. Most notably, the nature of the substitution relationships between the pest management inputs and other inputs are affected by this technological structure (see Berndt and Christensen, 1973). Separability also plays an important role in the analysis of pest management technology because it may allow pest management activities to be modeled and measured separately from other production activities. The role of separability is exploited in the case study presented in chapter 5.

Productivity, Policy, and Producer Welfare

The effects of pest management activities on overall technical and allocative efficiency and producer welfare can be evaluated by using the concepts outlined above. An increase in the technical efficiency of the pest management process means that more service flow x_m is obtained from a given input set (v_m, z_m). If x_m has a positive marginal product—that is, if $\partial f/\partial x_m > 0$—higher technical efficiency in pest management increases output, profits, and welfare. Similarly, the introduction of new or improved pest management technology changes the pest management function (3.5) in a manner such that higher output and producer welfare are obtained. In the terms of figure 3–1, an increase in the productive efficiency of pest management inputs causes the function J^* to shift upward to indicate a higher level of profit for a given input. If it can be assumed that x_m has a positive marginal product, and if pest management activities are separable, an increase in overall productivity is implied by an increase in pest management productivity.

Assuming the separability of pest management activities, the condition for allocative efficiency is

$$\frac{\partial q}{\partial x_m}\frac{\partial x_m}{\partial v_k} - w_k = 0 \qquad (3.6)$$

This condition shows that the efficient allocation of pest management inputs depends on both the marginal productivity of the input in the pest management process and the marginal productivity of pest management as an input into the overall production process. To evaluate the allocative efficiency of pest management inputs, it is necessary to take into account both the pest management subprocess and its role in the overall production process.

Technological change in pest management that is embodied in inputs such as pesticides should increase the marginal productivity of pest management inputs relative to other inputs, and thus encourage their use relative to other inputs. In the language introduced above, such technological change leads to a bias in the technology toward pesticides, in the sense that at given input levels the marginal productivity of pesticides is increased relative to other inputs. Consequently, the cost-minimizing farm manager would perceive an incentive to use more pesticides relative to other inputs, and the share of pesticides in total cost would increase. Indeed, in the post-World War II era in the United States, the data suggest that technological change has been biased toward chemical-intensive technology. It is likely that this has been a result in part of the bias of technological change toward both fertilizers and pesticides.

Also possible are other kinds of change in pest management technology which do not necessarily bias the technology toward pesticides. Research conducted under the IPM rubric may fall into this category. For example, much research has been devoted to the efficient use of chemical pesticides. Techniques such as the systematic sampling of fields for pests, which allows farmers to use pesticides efficiently and only as needed, may bias the farmer's technology toward the use of labor for field monitoring, and allow farmers to obtain higher yields with less pesticide input. The development of plant varieties that are resistant to pests also allows farmers to substitute an improved variety for pesticides, and thus represents a form of technological change that is biased *against* pesticides.

The effects on productivity and producer welfare of environmental policies which impose regulations on pesticide use can be analyzed by applying technical and allocative efficiency conditions. If a restriction on the use of a pesticide simply forces a producer to use less than the amount that would otherwise be chosen, the restriction would affect the producer's allocative efficiency, but would not necessarily affect technical efficiency. Technical efficiency would be affected by input regulations if they made the current pest management practices infeasible. In that case, input regulations could force a producer to use a pest management technology that is technically less efficient than the one employing the restricted pesticide.

Whether a policy restricting pesticide use would affect only allocative efficiency or both allocative and technical efficiency is an important question. If only allocative efficiency would be affected, estimates of the technology currently in use would provide the basis for evaluation of the policy's effects on producer welfare. However, if the policy would induce a change of technique, it would be necessary, in order to evaluate welfare effects of the policy, to measure producer welfare under both the pre- and post-policy technologies. To do so would present a serious empirical problem if the effects of the policy must be analyzed ex ante and there were no observations of farmers actually using the new technique.

If agricultural producers are allocatively efficient before pesticide regulations are instituted, a restriction on use of a pest management input v_k moves the producer away from efficiency, and the regulation necessarily reduces producer welfare. If the producer were underutilizing a pesticide from the allocative efficiency point of view (in figure 3–1, if the producer were to the left of point A), a binding restriction on its use would lower the producer's welfare. However, if the producer were overusing a pesticide, in the sense that the value of the marginal product was less than the normalized factor price (in figure 3–1, if the producer's input choice was greater than point A), a restriction on its use could either increase or decrease producer welfare, depending on where the restriction placed the producer relative to the initial welfare level. It can be concluded that the effects of pesticide regulation on allocative efficiency cannot be determined without prior determination of producers' allocative efficiencies.

One important implication of the neoclassical model is that, in general, inputs are substitutable for one another to a degree determined by the structure of the production technology. The substitutability issue is significant in the context of pesticide regulations because it determines the degree to which restrictions on the use of one particular pesticide, or on the use of pesticides in general, reduce productivity and producer welfare. For example, if pest control is in fixed proportions to one chemical, a restriction on that chemical would lead to a proportional reduction in pest management services and hence a reduction in overall productivity. However, if there were other means of controlling pests, either by substituting other chemicals or by nonchemical inputs, the regulation would have a smaller effect on the farmer's ability to produce pest management services.

Generalizations of the Neoclassical Model

The above discussion shows that the neoclassical model yields important insights into the analysis of pest management technology.

The concepts of technical and allocative efficiency and producer surplus provide an objective framework in which producer behavior and welfare can be measured and evaluated. The structure of the production technology, as regards input substitutability and separability, has implications for the measurement of pest management productivity and for the analysis of policies concerning such issues as the regulation of pesticide use and the introduction of new pest management technologies.

While the neoclassical model provides a framework within which many issues related to production and policy can be analyzed, it suffers two major shortcomings for the analysis of pest management technology. First, it is static—it does not reflect the time dimension of agricultural production processes. As noted in chapter 1, the sequential acquisition and utilization of information about pest populations is an essential element of efficient pesticide use, but this dimension of pesticide productivity cannot be captured in the neoclassical model. Second, the neoclassical model ignores the uncertainty inherent in agricultural production processes—uncertainty resulting from the unpredictability of weather, pest infestations, and other random events. Thus the neoclassical model fails to capture two essential dimensions of pest management activities.

To overcome these shortcomings of the neoclassical model, it can be generalized in several directions. The static, nonstochastic production function is extended to the dynamic, stochastic case. In this way sequential decision making and production risk can be introduced into the analysis of pest management technology. The profit maximization assumption is generalized to account for the stochastic technology, and corresponding efficiency and welfare measures are developed.

Modeling Stochastic Production Processes

The logical way to generalize the neoclassical theory of production to the stochastic case is to define the firm's output and prices as stochastic processes satisfying certain conditions that are analogous to the neoclassical regularity conditions, and which define the technological and economic conditions faced by the firm. In the fullest generality, outputs and prices may be described in terms of a joint probability distribution. To simplify the discussion that follows, firms are assumed to be price takers, and prices and outputs are assumed to be statistically independent. The latter assumption allows the firm's stochastic technology to be defined separately from price distributions, in a manner analogous to the definition in the neoclassical theory of the firm, in which the production function is distinct from prices. In the more general case in which output and prices are correlated, one must distinguish the output distribution conditional on prices from the marginal output distribution.

The neoclassical production function can be generalized to the stochastic case by adding a random term to the variables in the production function—that is, equation (3.1) is written as $q = q(x, z, \phi, \epsilon)$, where ϵ represents the randomness in output from factors beyond the firm's control, such as weather or mechanical breakdown. Alternatively, viewing output as a stochastic process generated by the firm's production decisions suggests that the production process can be represented as a probability distribution function. In the most general case, output q and inputs x and z can be interpreted as jointly distributed random variables, in a manner analogous to the most general representation of an implicit neoclassical production function. But for most purposes the stochastic production process can be represented as a stochastic process for output conditioned on the inputs—that is, as the conditional probability distribution $F[q|x, z, \phi]$.

Both the production function representation and the distribution function representation are useful and are discussed in this study. However, it should be noted that the production function interpretation generally imposes more restrictions on the stochastic structure of the production process than the conditional distribution specification does, because of the restrictions that the neoclassical theory imposes on the production function. In the next chapter it is shown that many commonly used specifications of the stochastic production function impose arbitrary restrictions on the stochastic structure of the production process.

The static neoclassical production model, as outlined in equations (3.1) through (3.4), can be generalized formally to the stochastic case. The firm's output q, its vector of fixed and variable inputs x and z, and output price p are defined on nonnegative, bounded subsets S_q, S_x, S_z, and S_p of Euclidean space. The firm formulates output price expectations based on an information set ω, which is defined on an appropriately dimensioned subset S_ω of Euclidean space. In the case of static production, the production process is defined in terms of the conditional probability distribution function $F[q|x, z, \phi]$: $S_q X S_x X S_z \rightarrow [0, 1]$. Alternatively, assuming that the probability distribution function is continuous and differentiable everywhere, the technology can be defined in terms of the corresponding probability density function

$$f(q|x, z, \phi) = \frac{d}{dq} F[q|x, z, \phi]. \tag{3.7}$$

(The convention will be maintained throughout that lowercase letters denote density functions and uppercase letters denote cumulative functions.) The distribution of output price is defined as $G[p|\omega]$: $S_p X S_\omega \rightarrow [0,1]$. In place of the neoclassical regularity conditions, it is assumed that the probability distribution function is continuous and twice differentiable for all $x \in S_x$, such that the choice problem

$$\max \int_{q \in S_q} \int_{p \in S_p} (pq - wx)\, dF[q|x, z, \phi]\, dG[p|\omega] \tag{3.8}$$

has a unique interior solution. The first-order condition for the solution of the maximization problem in equation (3.8) is

$$E(p)\frac{\partial E(q)}{\partial x} - w = 0,$$

where $E(\cdot)$ is the mathematical expectation operator. Assuming differentiation under the integral is valid, the first-order condition can be written

$$\left(\int p\, dG[p|\omega]\right)\left(\int q\, \partial f[q|x, z, \phi]/\partial x\, dq\right) - w = 0$$

This equation can be interpreted as showing that the firm equates the expected value of the marginal product to the factor price at the optimum. Note that the second set of terms on the left-hand side in the above equation can be interpreted as the expected marginal product of x. This set indicates how a change in x alters the conditional probability distribution of q.

If the firm is risk averse, that is if the firm's decision problem is to maximize the expected utility of profit and the utility function is concave, then $U(pq - wx)$ replaces $pq - wx$ in equation (3.8), where $U(\cdot)$ is the utility function satisfying $U' > 0$ and $U'' < 0$. The first-order condition for the optimal input choice is

$$\text{EMU} \equiv \partial EU[pq - wx]/\partial x = 0,$$

or

$$\iint U[pq - wx]\, \partial f[q|x, z, \phi]/\partial x\, dq\, dG[p|\omega]$$
$$- w \iint U'\, dF[q|x, z, \phi]\, dG[p|\omega] = 0$$

The first term in the above equation can be interpreted as the analog to the expectation of the value of the marginal product, and the second term is marginal factor cost in units of utility (note that the term multiplying the input price is expected marginal utility). Thus these equations define the marginal condition represented by the intersection of the EMU curve with the horizontal axis in figure 2–5. Under the assumption that the expected utility function is concave in x, the EMU curve is decreasing in x as drawn in the figure 2–5.

An alternative approach to the analysis of the firm under uncertainty is to use the stochastic production function to represent the firm's technology. Thus, define the stochastic production function as $q[x, z, \phi, \epsilon]$, where ϵ is a random variable with probability distribution function $\Phi(\epsilon)$. The expected utility of profit is then written

$$EU[\pi] = \int\int U[pq(x, z, \phi, \epsilon) - wx]\, d\Phi[\epsilon]\, dG[p|\omega]$$

and the first-order condition for expected utility maximization is

$$\int\int U'(p\partial q/\partial x - w)\, d\Phi[\epsilon]\, dG[p|\omega] = 0$$

Rearranging, and using the fact that $E(ab) = E(a)E(b) + \text{cov}(a, b)$, where $\text{cov}(a, b)$ is the covariance, gives

$$E(p)\partial E(q)/\partial x + E(p)\text{cov}(U', \partial q/\partial x)/E(U') = w$$

Note that if the covariance is zero, the first-order condition is the same as the one that is given for the risk-neutral farmer following equation (3.8). When the covariance is nonzero, the risk-averse farmer's input use differs from that of the risk-neutral farmer as a function of the second term on the left-hand side. Since this term measures the wedge that risk aversion drives between the expectation of the value of the marginal product and the factor price, it is equal to the marginal risk premium defined in chapter 2.

An important interpretation can be put on this first-order condition in terms of the kind of effect an input has on a risk-averse producer. It should be noted that a positive shock to production increases profit and thus moves the producer to a higher utility level. Assuming risk aversion, marginal utility U' declines as a result of such a shock. At the same time, the marginal product may be either increasing or decreasing. If the marginal product increases, then $\text{cov}(U', \partial q/\partial x) < 0$, and the risk-averse farmer will use less of the input than would a risk-neutral farmer—that is, according to the Pope-Kramer (1979) definition, the input is marginally risk-increasing. Conversely, if the covariance is positive, the input is marginally risk-decreasing. In the case of pesticides, a negative shock to production (a pest infestation) should simultaneously reduce profit, increase U', and increase the marginal productivity of pesticides; hence the covariance should be positive and pesticides should be a risk-reducing input.

A third approach to the analysis of the firm under uncertainty is to express the firm's objective function as a function of the moments of the random variables in the utility function. Under some conditions this can be done exactly; more generally, the approach can be used to approximate the objective function. There are actually two avenues to this kind of "moment-based" approximation. One is to express the expected utility function in general as

$$EU[\pi] = u[\mu_1, \mu_2, \ldots, \mu_m],$$

where the μ_i are the moments of profit and depend on the conditioning variables x and z. The first-order conditions for expected utility maximization are

$$\partial EU/\partial x = \sum_{i=1}^{m} (\partial u/\partial \mu_i)(\partial \mu_i/\partial x) = 0$$

Noting that the derivative of expected profit with respect to x equals the expected value of marginal product minus the factor price, it can be seen that this equation can be rewritten as

$$E(p)\partial E(q)/\partial x + \sum_{i=2}^{m} r_i(\partial \mu_i/\partial x) = w,$$

where

$$r_i = (\partial u/\partial \mu_i)/(\partial u/\partial \mu_1)$$

Thus the term involving the summation in the above equation can be interpreted as the marginal risk premium, as it indicates the wedge that risk drives between the expected value of marginal product and the factor price. It follows that if none of the higher moments are functions of x, the first-order condition of the risk-averse firm coincides with the first-order condition of the risk-neutral firm.

It also is possible to use a Taylor series to approximate the objective function. Taking its expectation, one obtains expressions in terms of the moments of the random variables. Both of these versions of the moment-based approach are used in the econometric developments in chapter 4 to translate the theoretical model of the firm under uncertainty into an empirically useful model.

Production Dynamics and Sequential Decision Making

The agricultural production process occurs over long time periods and is integrally related to the biological processes involved. Production decisions must take into consideration the biological growth of each crop and the interrelationships among crops. Thus it is useful to divide production decisions into two categories, intraseasonal and interseasonal. The former refer to decisions made with respect to a sequence of intermediate production stages that lead to a final output in a growing season; the latter refer to decisions related to a sequence of final outputs across seasons. This study focuses on intraseasonal decisions associated with intermediate input decisions such as pesticide and fertilizer applications.

Intraseasonal decisions typically relate to multistage production processes; that is, they involve a sequence of inputs applied over time in a sequence of intermediate production stages. These intermediate stages lead to a final output. In this case, the production process can be said to exhibit *output dynamics*, and can be defined formally as a stochastic process of the form $F_t[q_t|x_t, q^{t-1}, \phi]$, where the vector $q^{t-1} = (q_{t-k}, \ldots, q_{t-2}, q_{t-1})$ contains past output realizations that affect q_t (generally, the number k of lagged outputs conditioning the distribution may vary with t). Thus in the case of output dynamics, the distribution of q_t depends on inputs x_t and on the past sequence of output realizations.

Within a given production process or stage of production, there may be either a single input or a sequence of inputs over time. When there is a sequence of inputs for each output, the production process can be said to exhibit *input dynamics*, and the distribution of output at time t is $F_t[q_t|x^t, \phi]$, where $x^t = (x_{t-j}, \ldots, x_{t-1}, x_t)$. In general, the production process may exhibit both input and output dynamics, in which case the sequence of outputs $\{q_t\}$ is generated by $F_t[q_t|x^t, q^{t-1}, \phi]$. Note that the distribution function is dated, to reflect the fact that each one represents a distinct process or stage of production.

In the dynamic version of the stochastic model (3.8), the risk-neutral firm chooses the *sequence* of variable inputs $\{x_t\}$ to maximize the present value of profit, given its technology and its information set Ω_t in each period. Generally, the information set Ω_t contains some subset of the relevant histories x^{t-1}, q^{t-1}, r^{t-1}, and p^{t-1}, as well as the discount factor, and other relevant economic or technological information such as the parameters of the relevant price and output distributions. Within a given season with $t = 1, \ldots, T$ stages of production, a risk-neutral firm whose production process exhibits output dynamics has as its objective the choice of input sequence $\{x_1, \ldots, x_T\}$ to maximize

$$E_0[J|\Omega_0] = \int \ldots \int \left[p_T q_T - \sum_{t=1}^{T} w_t x_t \right] dF_T[q_T|x_T, q_{T-1}] \ldots dF_1[q_1|x_1]\, dG[p_T|\omega], \tag{3.9}$$

where p_T is the price of final output (for simplicity, assume that the distribution of p_T is not updated over the season, hence its distribution G is not dated), input prices w_t are assumed to be known and non-stochastic, and E_0 represents the expectation taken with respect to random variables at time 0. More generally, the short-run decisions within each season are nested within a longer-run set of interseasonal decisions. Since the concern here is with the intraseasonal decisions related to pest management, the longer-run case is not explored (for that case, see Antle, 1988).

The general solution to maximization of equation (3.9) can be obtained from the dynamic programming algorithm (see Aoki, 1967). The dynamic programming approach is to solve the maximization problem recursively, solving the choice problem in period t conditional on decisions in periods $1, \ldots, t-1$, and on the optimal decision rules for future periods $t+1, \ldots, T$. That is, the solution is obtained by solving, for each t,

$$\max_{x_t} E_t[J_t | \Omega_t],$$

where

$$\Omega_0 = (\omega, w_1, \ldots, w_T), \qquad \Omega_t = (\Omega_0, \ldots, \Omega_{t-1}, x^*_{t+1}, \ldots, x^*_T),$$

and x^*_t is the optimal decision rule for x_t.

To illustrate the sequential decision-making process involved with this kind of dynamic production problem, consider a production process having three production stages. These can represent crop-growth stages and corresponding sequential operations within a growing season, such as planting, cultivation, and harvesting. The stages also can represent pest management decisions over the growing season. The sequence of events in the decision-making process is as follows:

In stage 1 the variable input x_1 and fixed factor z are chosen, based on initial expectations of prices, future crop states, and decision rules for future inputs x_2 and x_3, given by

$$\begin{aligned} x^*_2 &= x^*_2(w_2, w_3, q_1, \omega) \\ x^*_3 &= x^*_3(w_2, w_3, q_1, q_2, \omega) \end{aligned} \qquad (3.10)$$

Thus in stage 1 the firm chooses x_1 to solve

$$\max_{x_1} E_1[J | \Omega_1] \quad \text{where } \Omega_1 = (x^*_2, x^*_3, \Omega_0),$$

$$J = p_3 q_3 - w_1 x_1 - w_2 x_2 - w_3 x_3,$$

and E_t denotes the mathematical expectation operator taken with respect to the random variables at time t. After x_1 is chosen, production begins, and q_1 is realized.

At the beginning of stage 2 the farmer solves

$$\max_{x_2} E_2[J|\Omega_2], \quad \Omega_2 = (q_1, \Omega_1),$$

production in stage 2 then begins, and q_2 is realized.

In stage 3 the farmer solves

$$\max_{x_3} E_3[J|\omega], \quad \Omega_3 = (q_1, q_2, \Omega_2)$$

An important feature of the optimal solution of the dynamic decision problem is the updating of the information set Ω_t over time. It is this feature of the decision-making process which causes the optimal factor demands given by (3.10) to depend on the output realizations over time. This property of the factor demand functions in the sequential decision process has important implications for the econometric measurement of pest management technologies discussed in chapter 4.

The sequential input choices made in the dynamic, stochastic model differ from the factor demands of the neoclassical model in terms of the first-order condition that is satisfied at the optimum. In the neoclassical model, the value of the marginal product is equated to the factor price, as in equation (3.3), and factor demand functions depend on all variable factor prices, output price, and fixed factors. In the dynamic model, the sequential input demand functions depend on previous variable input quantities, output price expectations, and expected input quantities in future periods. Thus in the dynamic model previous decisions, uncertainty, and the effect of current decisions on future decisions must be taken into account. To illustrate, consider the solution of the stage 2 problem outlined above. The first-order condition is

$$\frac{dE_2[p_3 q_3|\Omega_2]}{dx_2} = \frac{dE_2[\Sigma\, w_t x_t|\Omega_2]}{dx_2}$$

or

$$E_2[p_3|\Omega_2]E_2\left[\frac{\partial q_3}{\partial x_2} + \frac{\partial q_3}{\partial x_3}\frac{\partial x_3}{\partial x_2}\middle|\Omega_2\right] = w_2 + \frac{\partial E_2}{\partial x_2}[w_3 x_3^*|\Omega_2]$$

These equations state that the optimal choice for x_2 equates the expected value marginal product (EVMP) on the left-hand side to the expected marginal factor cost (EMFC) on the right-hand side. EVMP includes the direct effect of x_2 on output through the production process, plus the indirect effects of x_2 on output due to its effect on the choice of x_3. Similarly, EMFC includes the factor payment w_2 plus the effect on cost in the next stage.

Separability in Stochastic Technologies

The discussion of the neoclassical model has suggested that separability is an important property of agricultural technologies. To extend the concept of separability to stochastic technology, consider a production process for output q with two subprocesses for inputs x_g (general production inputs) and x_m (pest management service flows). Suppose further that each of the subprocesses generating x_g and x_m are stochastic and are jointly distributed according to $F[x_g, x_m | v_g, v_m, z_g, z_m]$, where v_i denotes a variable input vector and z_i a fixed input vector, $i = g, m$. For simplicity, suppose further that q is related to x_m and x_g according to the relation

$$q = q(x_m, x_g) \tag{3.11}$$

Thus q is a random variable because of the randomness of x_g and x_m. The joint probability density of q can be written

$$f_q[q | v_g, v_m, z_g, z_m] = (1/2\pi) \int \exp(-itq)\, \chi(t)\, dt, \tag{3.12}$$

where

$$\chi(t) = \iint \exp(itq)\, dF[x_m, x_g | v_g, v_m, z_g, z_m] \tag{3.13}$$

Expression (3.13) is known as the *characteristic function*, and uniquely determines the distribution of q (Kendall and Stuart, 1977). The relation of x_g and x_m to q is complicated, unless structure such as additivity of (3.11) in x_m and x_g is assumed.

Equations (3.11) through (3.13) show that, in general, production inputs are nonseparable in the sense that there is no distinct subfunction which relates the service flows x_g and x_m to a separable subprocess. Such separability in the stochastic case requires that two conditions be met. First, the inputs into the subprocess must affect only the marginal distribution of the associated service flow. Second, the variables x_g and x_m must be statistically independent. Under these two conditions, the joint distribution of x_g and x_m can be factored as

$$f[x_g, x_m | v_g, v_m, z_g, z_m] = f_g[x_g | v_g, z_g] \, f_m[x_m | v_m, z_m]$$

It follows that the distribution of q can be written

$$f_q[q | v_g, v_m, z_g, z_m] = \tilde{f}\,[\, f_g[x_g | v_g, z_g],\, f_m[x_m | v_m, z_m]] \tag{3.14}$$

When expression (3.14) is satisfied the stochastic technology is said to exhibit stochastic separability. This condition is analogous to functional separability in the neoclassical model. A stronger version of stochastic separability (which could be referred to as strong stochastic separability) can be defined as the condition in which (3.11) is additive in x_m and x_g. In that case, it can be shown (Freund, 1972) that

$$f_q[q | v_g, v_m, z_g, z_m] = \int f_g[q - x_m | v_g, z_g] \, f_m[x_m | v_m, z_m] \, dx_m$$

In the context of pest management, stochastic separability implies that the pest management inputs such as pesticides do not directly affect other dimensions of the biological processes involved in plant growth. It is a well-established fact, however, that in some cases pesticides do have significant effects on plant growth. Thus one must give careful consideration to these effects in applied research. However, in attempting to obtain a good approximation to the actual production process, if such growth effects are of second or smaller order of importance compared to the pest management effects, the assumption of separability may be desirable as a significant simplification of a highly complex process.

Efficiency and Producer Welfare Under Uncertainty

As we have seen, the neoclassical analysis of producer welfare is based on producer surplus. Since the firm is assumed to make production decisions so as to maximize profit, it maximizes producer surplus and thus its welfare. Under uncertainty, the firm's objective is to maximize expected utility or some other function of the technological and behavioral parameters. Thus under uncertainty the logical generalization of the neoclassical concepts of efficiency and welfare are based on the firm's objective function.

Welfare under uncertainty can be measured in terms of the units in which the firm's objective function is defined, such as units of utility, or that measurement can be made by translating the units into a numeraire such as dollars. The standard practice in welfare economics is to do the latter. For example, Pope, Chavas, and Just (1983) consider the firm with the objective function

$$\max E[U(W + \pi)],$$

where W is initial wealth, and π is a random variable such as profits or producer surplus. The firm's risk premium is defined as R. R is the maximum amount of money the firm would be willing to pay to obtain expected profit instead of facing the risk associated with profit. In other words, R is the maximum amount the firm would be willing to pay to insure against the risk associated with π. It follows that R satisfies

$$U(W + E[\pi] - R) = E[U(W + \pi)]$$

That is, the utility of obtaining $W + E[\pi]$ with certainty, less the insurance cost R, equals the utility of the risky prospect π. The amount $W + E[\pi] - R$ that the firm could obtain with certainty if it were insured against the risk of π is defined as the *certainty equivalent*, and is equal to

$$CE \equiv W + E[\pi] - R = U^{-1}(E[U(W + \pi)])$$

The firm's certainty equivalent thus is a measure of welfare in dollar terms, and choosing x to maximize CE is equivalent to maximizing $E[U(\pi)]$, as long as $U' > 0$.

The first-order condition for maximizing CE is

$$\partial CE/\partial x = \partial E[\pi]/\partial x - \partial R/\partial x = 0$$

Note that this equation involves the marginal change in expected profit and the marginal risk premium, and therefore is identical to the first-order conditions derived directly from the utility maximization models discussed following equation (2.8).

As already noted, if the risk premium is not a function of x, the input choice problem of the risk-averse firm is equivalent to that of the risk-neutral firm. The area under the expected marginal value product (EVMP) curve then is a valid measure of welfare for both risk-averse and risk-neutral firms. To better understand why this is true, consider the following simple example. Let the utility function be the negative exponential function $U(\pi) = 1 - \exp(-2\gamma\pi)$, and assume that π is normally distributed. Taking the expectation of $U(\pi)$ and solving for CE, it follows that

$$CE = \mu_1 - \gamma\mu_2,$$

where the first term is mean profit and the second term is the risk premium. The risk premium in this case equals the product of the absolute risk aversion coefficient γ and the variance of profit μ_2. Thus if the risk premium does not change with x, there must be constant absolute risk aversion (that is, γ is a fixed parameter that does not vary with wealth or income), and the higher moments of the distribution of π must not be functions of x.

In the more likely case that risk aversion is not constant or x affects production risk, $\partial R/\partial x \neq 0$ and the conventional welfare measure based on EVMP is not valid. This occurs when preferences exhibit decreasing or increasing absolute risk aversion, because then a change in profit induces a change in risk aversion, or higher moments of the profit distribution are affected by inputs. In these more typical cases, a valid welfare measure must take into account the effects of input use on the risk premium.

Welfare can be measured in terms of money (the area under an input demand curve), or directly in terms of utility. Because preferences are not directly observable, the conventional approach to welfare measurement is to compute money welfare measures by integrating areas above or below demand or supply functions. This approach involves various well-known theoretical and practical problems. Theory tells us that the areas under demand or supply curves measure utility if, and only if, they are the true "compensated" (that is, utility constant) curves. Research has shown, however, that the error introduced by using uncompensated demand or supply functions is typically small (Willig, 1976; Pope and Chavas, 1985). In considering the errors of measurement involved in empirical research, then, the distinction between compensated and uncompensated demand functions may not be important. But even if this distinction is ignored, it is still necessary to estimate the risk-adjusted supply or demand functions which satisfy certain theoretical properties (integrability, for example) in order to measure welfare as an area associated with them. While this is possible in principle, it generally involves difficult estimation problems.

If the utility function is known one can simply compute welfare changes in utility terms, or transform the utility change into money terms, using the inverse utility function. This is the approach pursued in the present study. In chapter 4 it is shown that the producers' risk attitudes, and hence their utility parameters, can be estimated by using established econometric methods. Thus by assuming a functional form for a representative utility function, welfare analysis can be implemented.

It is observed in chapter 2 that the conventional concept of economic efficiency can be related directly to producer welfare: a reduction in either technical or allocative efficiency implies a reduction in economic efficiency, measured as a change in producer surplus. And it has been noted in the discussion of the neoclassical model that the welfare effects of policies may depend on the degree of allocative or technical efficiency of producers. For policy analysis, then, it may be useful to assess producers' efficiencies, as well as to decompose welfare changes into components attributable to technical and allocative efficiency changes. The following two sections therefore develop efficiency concepts that are appropriate under uncertainty, and efficiency indexes that are analogous to the neoclassical efficiency indexes.

Assumptions and Definitions for Efficiency Analysis

A neoclassical technology can be interpreted as a frontier production function showing the maximum possible output for a given input set and technology, or as a set of related production functions of varying degrees of efficiency. The latter interpretation is used here to define a stochastic technology as a set $T = \{f(q|x, z, \phi): x, z \in X_T\}$ of conditional output distribution functions associated with a given set X_T of variable and fixed inputs. For the purposes of this discussion, the set T is defined as a finite, discrete set of distributions; more generally it could be defined over a closed, compact set of attributes in Euclidean space. Note that an alternative definition of a stochastic technology could be made in terms of the corresponding set of revenue or profit distributions.

The firm's objective function is defined in terms of the technology, behavioral parameters, and input choices. For a given vector of behavioral parameters α, input vectors $x, z \in X_T$, and a production process $f^i(q|x, z, \phi) \in T$, the firm is assumed to possess a real-valued scalar objective function $J^i(x, z, \alpha)$ which is globally concave in x and has a unique interior global maximum at x^*. The vector α represents all variables related to the firm's objectives that are not elements of x or z, including the technology parameters ϕ, prices, risk attitudes, and wealth, for example.

Using the above definition of a stochastic technology and the firm's objective function, the following definitions of technical and allocative efficiency under uncertainty are proposed:

DEFINITION 1. A production process $f^*(q|x,z,\phi) \in T$ is *absolutely technically efficient* if, and only if, $J^*(x,z,\alpha) = \max \{J^i(x,z,\alpha) : f^i(q|x,z,\phi) \in T; x,z \in X_T\}$.

DEFINITION 2. For any subset T_i of T, a production process $f^i(q|x,z,\phi)\epsilon T_i$ is *relatively technically efficient* if, and only if, $J^i(x,z,\alpha) = \max\{J^j(x,z,\alpha): f^j(q|x,z,\phi)\epsilon T; x,z\epsilon X_{T_i}\}$.

DEFINITION 3. An input choice $x_k \epsilon T$ (an element of vector x) is *absolutely allocatively efficient* for $f^* \epsilon T$ if, and only if, it satisfies $\partial J^*(x,z,\alpha)/\partial x_k = 0$.

DEFINITION 4. An input choice $x_k \epsilon X_{T_i}$ is *relatively allocatively efficient* for $f^i \epsilon T$ if, and only if, it satisfies $\partial J^i(x,z,\alpha)/\partial x_k = 0$.

A production process f^* is absolutely efficient, according to definition 1, if and only if it yields the maximum feasible value of the firm's objective function $J(x,z,\alpha)$ for a given technology T, input set $x,z \epsilon X_T$, and parameter vector α. Thus f^* is analogous to the frontier production function in neoclassical theory. Similarly, the definitions of relative efficiency under uncertainty are analogous to the relative neoclassical efficiency concepts discussed earlier.

The preceding definitions state that the value of the firm's objective function depends on the input vectors x and z at which it is evaluated; hence, the firm's degree of technical efficiency generally depends on x and z. Technical efficiency also generally depends on factor proportions in the neoclassical model. But in some cases the efficiency ordering may be independent of inputs. For analysis of these relationships the following definition is useful:

DEFINITION 5. A production process $f(q|x,z,\phi)\epsilon T$ is *uniformly efficient* if, and only if, it is technically efficient for all $x,z \epsilon X_T$.

Efficiency Indexes

Technical and allocative efficiency under uncertainty can be interpreted graphically by using figure 3–1. Uniform efficiency is shown in that figure as $J^* > J^i$ for all input levels. J^* is interpreted as the objective function for the absolutely efficient production process, and J^i is interpreted as the objective function for the relatively efficient process f^i. The points in the figure are interpreted as follows:

A) $J^*(x^*,z,\alpha)$, absolute technical and allocative efficiency;
B) $J^*(x^2,z,\alpha)$, absolute technical efficiency; absolute and relative allocative inefficiency;
C) $J^i(x^1,z,\alpha)$, absolute technical inefficiency; relative technical and allocative efficiency;

D) $J^i(x^2, z, \alpha)$, absolute technical and allocative inefficiency; relative technical efficiency; relative allocative inefficiency.

Figure 3–1 suggests a method of measuring technical and allocative efficiency at any point, provided the efficient and inefficient technologies are known. Suppose the firm is observed producing with input x^2 and technology $f^i(q|x, z, \phi)$. The absolute total welfare loss to the firm, or total loss in economic efficiency, resulting from technical and allocative inefficiency at point x^2 can be measured as $J^*(x^*, z, \alpha) - J^i(x^2, z, \alpha)$. This loss can be divided into an absolute allocative inefficiency component $J^*(x^*, z, \alpha) - J^*(x^2, z, \alpha)$, attributable to the fact that the firm is producing with an allocatively inefficient input quantity, and an absolute technical inefficiency component $J^*(x^2, z, \alpha) - J^i(x^2, z, \alpha)$, because the firm is producing with an absolutely inefficient production process. These measures of welfare loss can be related to indexes of absolute technical efficiency (ATE), absolute allocative efficiency (AAE), and absolute economic efficiency (AEE) as follows:

$$\begin{aligned} \text{ATE} &= J^i(x^2, z, \alpha)/J^*(x^2, z, \alpha) \\ \text{AAE} &= J^*(x^2, z, \alpha)/J^*(x^*, z, \alpha) \\ \text{AEE} &= J^i(x^2, z, \alpha)/J^*(x^*, z, \alpha) = (\text{ATE})(\text{AAE}) \end{aligned} \tag{3.15}$$

Assuming that the objective function's origin is at zero, as in figure 3–1, and that f^* is uniformly efficient, these indexes range from zero to one. At a given input level x^2, an increase in technical efficiency implies higher values of ATE and AEE; an increase in allocative efficiency implies higher values of AAE and AEE.

The indexes in equation (3.15) are based on the assumption of a uniformly efficient production process for all x. If there is no uniformly efficient process, the efficient objective functions "cross" at one or more points in input space. In this case the efficiency indexes in (3.15) are well defined on each set X_i for which process f_i is efficient, but cannot be used to compare efficiency across sets.

A related set of relative efficiency measures also can be defined. For any technology $f^j \epsilon T_i$ (see definition 3) and its corresponding objective function J^j,

$$\begin{aligned} \text{RTE} &= J^j(x^2, z, \alpha)/J^i(x^2, z, \alpha) \\ \text{RAE} &= J^i(x^2, z, \alpha)/J^i(x^1, z, \alpha) \\ \text{REE} &= J^j(x^2, z, \alpha)/J^i(x^1, z, \alpha) = (\text{RTE})(\text{RAE}) \end{aligned} \tag{3.16}$$

Note that these efficiency indexes measure the firm's deviation from its optimum as the change in the value of its objective function. These measures relate to the vertical distance between the points in figure 3–1. A variety of other efficiency measures can be defined using other measures of distance from the optimum. For example, a measure could be defined as the horizontal distance (that is, deviations of the actual input level from the optimum), or as the slope of the function from the actual to the optimum. These three types of measure of distance all have precedents in the statistics literature on hypothesis testing. The vertical measure corresponds to the likelihood ratio test; the horizontal measure corresponds to the Wald test; and the slope measure corresponds to the Lagrange multiplier test (see Engle, 1984). The motivation for using the vertical measure in the context of efficiency analysis is that it has an interpretation in terms of welfare loss.

Another way to measure efficiency is to translate the utility measure of welfare into certainty equivalents. This would produce a set of indexes having the same qualitative properties as those defined above, although the numerical values of the certainty equivalent indexes generally would differ from the utility indexes.

The indexes in (3.15) and (3.16) have several notable properties, and can be compared with efficiency indexes for the neoclassical model. First, since inefficiency is measured as the sum of the firm's welfare losses resulting from its allocative and technical inefficiencies, the indexes have the multiplicative property AEE = (AAE)(ATE) and REE = (RAE)(RTE). In this respect these measures are similar to the Farrell measures. Second, the indexes defined in (3.15) and (3.16) depend on the parameter vector α. This means that in the Arrow-Pratt risk aversion model, for example, the firm's risk attitudes as embodied in the utility function parameters affect the measured degree of allocative, technical, and economic efficiency under uncertainty. It also means that these indexes generally are not invariant to prices, in contrast to the Farrell and other price-independent efficiency measures based on the neoclassical model. It can be shown, however, that under certain conditions the technical efficiency ordering is invariant to prices and other variables represented by α (Antle, 1985). Third, these indexes can be used to analyze multifactor or multiproduct technologies. The analysis of a scalar input illustrated in figure 3–1 is equally valid for the case in which x is interpreted as an input vector, and output is either a scalar or a vector.

A final property of these indexes is that they are defined on the (0,1) interval and can be interpreted as cardinal efficiency indexes only if the function J has a zero origin and is defined cardinally. If J is defined ordinally, the scaling of the function J is arbitrary, and the indexes have only an ordinal interpretation. For example, if J is interpreted as a cardinal von Neuman-Morgenstern expected utility function, it is uniquely determined up to a linear transformation. For positive numbers a and b, any other function $J^{\sim} = a + bJ$ is equivalent. This means, for example, that

$$\text{ATE}^{\sim} = J^{i\sim}/J^{*\sim} = (a + bJ^{i})/(a + bJ^{*}) \neq \text{ATE}$$

These indexes are invariant to the scale factor b if the origin a is zero. In making quantitative efficiency comparisons using indexes (3.15) or (3.16), only indexes calculated with the same origin can be compared. If the origin is nonzero, the indexes can be used to order efficiency, but the percentage of efficiency measured is a function of the choice of origin, and the index is not defined on $(0, 1)$. Similarly, if J is defined ordinally, the efficiency measures are less than or equal to unity, but an index of value .8 cannot be said to represent twice the efficiency of an index of value .4. All that can be inferred from the ordinal measure is a qualitative ranking.

The objective function J need not be interpreted as expected utility. A more general model can be defined in terms of profits and the distribution of profits, as discussed in Machina (1982). Other criteria that can be expressed as a maximization problem, such as a safety-first criterion, could be used (see Anderson, 1979).

The efficiency analysis as defined here also can accommodate dynamic models. For example, the objective function $J(x, z, \alpha)$ can be defined to be consistent with the sequential maximization of expected profit or utility, along the lines discussed earlier in this chapter.

Technological Change

The principles introduced for the analysis of efficiency under uncertainty also can be used to interpret technological change in the stochastic model. In the neoclassical model technological change is interpreted as a shift in the production function that enables the firm to obtain more output per unit of input or per unit of cost. When output is a random variable described by a conditional distribution, technological change can be interpreted as a change in that conditional distribution or in the inputs that condition the output distribution. Intuitively, technological progress in the stochastic case should mean that the process generating output changes in such a way that the firm is made better off, holding prices constant. As in the analysis of technical efficiency under uncertainty, a technologically superior production process therefore must provide the firm with a higher value for its objective function.

In the static risk-neutral case represented by (3.8), for example, a new production process F^1 that is economically superior to the old process F^0 should satisfy nonretrogression in the mean; that is, at constant factor prices, cost per unit of expected output should be nonincreasing:

$$wx/E_1[q|x,z] \leq wx/E_0[q|x,z]$$

However, in the more general case in which the objective function is more complex (in sequential decision making or risk aversion, for example), consideration of the mean alone is not sufficient to infer technological progress, for two reasons. First, the firm's objective function is nonlinear in output (even if the firm is risk neutral and decisions are made sequentially; see Antle, 1984). Second, the general representation of the production process as a stochastic process suggests that technological progress may involve beneficial changes in higher moments of the output distribution. The analysis of technological change is a multidimensional problem involving the mean as well as higher moments of output or profit.

These considerations suggest that the evaluation of technological change, like the analysis of technical efficiency, must involve the aggregation of the characteristics of the relevant distribution. It is evident from the preceding section on efficiency indexes that the logical basis for such aggregation is the firm's objective function. Thus the following definition of technological change is proposed:

DEFINITION 6. For two stochastic technologies T_0 and T_1, T_1 represents technological progress relative to T_0 if, and only if, $J^{*0}(x,z,\alpha) \leq J^{*1}(x,z,\alpha)$, where J^{*i} represents the maximum objective function value obtainable for technology T_i.

In other words, technological progress occurs when efficient use of a new stochastic technology would lead to an increase in the firm's welfare. Like the definitions of technical efficiency under uncertainty (definitions 1 and 2), technological progress generally must be evaluated for given values of the firm's inputs and parameters α. If technological progress occurs at all input levels, it can be said to be uniform in a manner analogous to definition 5.

Definition 6 suggests that the rate of technological change under uncertainty can be measured in a manner analogous to the measurement of absolute technical efficiency. The following definition is therefore suggested:

DEFINITION 7. The rate of disembodied technological change under uncertainty is

$$\rho(1,0) = \{J^{*1}(x,z,\alpha) - J^{*0}(x,z,\alpha)\}/J^{*1}(x,z,\alpha)$$
$$= 1 - \text{ATE}(J^{*1}, J^{*0})$$

In words, the rate of technological change is one minus the degree of absolute technical efficiency measured between the old and new technologies. As noted in the discussion of efficiency indexes, their interpretation depends on the type of objective function defined. A unique value of ρ is obtained within the class of objective functions with zero origin and arbitrary scaling. However, when the objective function has a nonzero origin or is ordinal, the index of ATE, and thus the measured rate of technological change, can be used only for ordinal comparisons. For example, for a given utility scaling, suppose that $\rho(1,0) = .01$ and $\rho(2,0) = .02$. In this case technology 2 generates a higher rate of technological change than technology 1; but it cannot be said that technology 2 represents a rate of technological change twice as great as technology 1, since with a different utility scaling $\rho(2,0)$ would be greater than $\rho(1,0)$, although it need not be exactly twice as great.

Another important concept in the analysis of technological change is the bias in that change (see Antle and Capalbo, 1988, for an extensive discussion of this bias and its measurement in neoclassical models). The bias in technological change was defined by Hicks in terms of the effect that technological change has on the marginal rate of technical substitution between two inputs. This concept can be generalized to the case of production uncertainty as follows. Consider the static maximization problem in equation (3.8). From its first-order conditions we can define

$$\text{SMRTS}_{i,j} \equiv \frac{\iint U(pq - wx)\, \partial dF[q|x,z,\phi]/\partial x_i \, dG[p|\omega]}{\iint U(pq - wx)\, \partial dF[q|x,z,\phi]/\partial x_j \, dG[p|\omega]} = \frac{w_i}{w_j}$$

$$= \frac{E(p)E(\partial q/\partial x_i) + E(p)\text{cov}(U', \partial q/\partial x_i)/E(U')}{E(p)E(\partial q/\partial x_j) + E(p)\text{cov}(U', \partial q/\partial x_j)/E(U')} = \frac{w_i}{w_j}$$

SMRTS is the stochastic marginal rate of substitution, the generalization of the marginal rate of technical substitution defined in the neoclassical model. Following the Hicksian concept of bias in that model, the bias in technological change can be defined as follows:

DEFINITION 8. Technological change is factor-i using (saving) as

$$B_{ij}(1,0) \equiv (\text{SMRTS}^1_{i,j} - \text{SMRTS}^0_{i,j})/\text{SMRTS}^0_{i,j}$$

is greater (less) than zero, where $\text{SMRTS}^k_{i,j}$ denotes the stochastic marginal rate of technical substitution between inputs i and j for technology k.

According to this definition, technological change is factor-i using if $\text{SMRTS}_{i,j}$ is increased by technological change, and factor-i saving if it is decreased. Note that the SMRTS depends on the expected value marginal product and on the marginal risk premium associated with each input. Thus the firm's risk attitudes play a role in determining the value of new technology to the firm: for a given change in mean marginal products, the risk-averse firm values the new technology more than the risk-neutral firm if the new technology reduces the degree of production risk. The bias in technological change under uncertainty depends on the way the technology changes both of these attributes of the firm's objective function.

Notes on the Literature

This chapter draws on the author's recent work, including Antle (1983a,b, 1985, 1986, 1987), and Antle and Park (1985). The literature on stochastic production functions and their relation to econometrics has its origins in the econometric literature on production function estimation; see Marschak and Andrews (1944), Mundlak and Hoch (1965), and Zellner, Kmenta, and Dreze (1966). The representation of yield distributions in terms of conditional probability distributions was introduced by Day (1965), and discussed further by Anderson (1973) and Roumasset (1976); see also Anderson, Dillon, and Hardaker (1977). Antle (1983b) makes the case for the use of the conditional distribution of output as a general representaton of stochastic technologies.

Production has long been recognized as a dynamic phenomenon in the economics and agricultural economics literatures. Early studies of investment behavior (Jorgenson, 1963; Lucas, 1967) introduced equations of motion for capital stocks and the concept of adjustment costs into the production literature. Nerlove's (1958) use of adaptive expectations to model agricultural supply response was another important contribution. More recently, dynamic optimization models have been used to explicitly introduce dynamics into production models (for example, Hansen and Sargent, 1980; Epstein and Yatchew,

1985). For a review of the literature, see Berndt, Morrison, and Watkins (1983).

The literature on the welfare analysis of the firm under uncertainty is rather recent. A series of papers by Chavas and Pope (1981), Pope, Chavas, and Just (1983), and Pope and Chavas (1985) establishes the foundations of the approach that is based on the producer surplus concept and the analysis of its validity in the case of price and production certainty. The use of direct-money equivalent, or money-metric, measures of utility based on preferences or the equivalent variation is discussed by McKenzie (1983).

four

ECONOMETRIC MEASUREMENT OF PRODUCER EFFICIENCY AND WELFARE

The preceding chapter identified two essential elements in the analysis of producer welfare: the farmer's dynamic, stochastic production technology and the farmer's objective function. This chapter develops an econometric methodology for quantifying both the stochastic technology and the decision-maker's objective function so that the theoretical concepts discussed in chapter 3 can be implemented empirically.

The methods presented here utilize moment-based approximations to the technology and the objective function. The chapter is divided into three main sections. The first provides the theoretical foundations for the moment-based approach to the analysis of the stochastic technology and describes econometric procedures for implementation of this approach, taking into account the effects of sequential input decisions on the econometric properties of the model. The second section shows that, given estimates of the technology, and assuming that a farmer's objective is the maximization of expected utility, the qualitative and quantitative attributes of the distribution of risk attitudes in the population of farmers can be inferred econometrically. The concluding section shows how the moment-based approach can be used to obtain a tractable approximation to the farmer's technology and objective function that can be used to analyze the productivity of pest management technology.

Measuring Stochastic Technology

The literature presents the research economist with a variety of methods for the measurement of stochastic technology. Early research based on the "method of moments" utilized experimental data (see

Kendall and Stuart, 1977, vol. 1, for a general discussion; see Day, 1965, Anderson, 1973, and Roumasset, 1976 for applications to production analysis). One major disadvantage of this approach is that it requires a sufficiently long time-series on each individual in the sample to obtain reliable estimates of the moments; such pooled time-series and cross-section data generally are not available for econometric analysis.

Econometric production function models also have been used to characterize the distribution of output and to estimate the moments of the distribution (for example, de Janvry, 1972; Just and Pope, 1978). Because these models are formulated by appending error terms to neoclassical production functions, they generally restrict the stochastic structure of the technology, and therefore may not be appropriate for the analysis of risk relationships. It will be shown below that these restrictions are particularly serious in the measurement of pest management technology. Another approach is to fit distribution curves to data and thus approximate the conditional distribution of output (see Taylor, 1984, for example). This type of approach has the advantage of flexibility, and can be used with nonexperimental data, but does not yield convenient algebraic expressions for factor demands or the measures of efficiency discussed in chapter 3.

The Moment-Based Approach

The flexible moment-based approach (which is to be distinguished from the method of moments) to measuring stochastic technology, as developed here, is motivated by the fact that under general conditions the probability distribution of output (or revenue or profits) is a unique function of its moments, and therefore the behavior of the firm can be defined in terms of the relationships between inputs and these moments. Any characteristic of a firm's stochastic technology can be measured and tested by using the moment functions that define the distribution. The moment-based approach, as essentially a generalized regression method, has the advantage of being feasibly applicable to a single cross-section of data, or with any number of pooled time-series and cross-section observations. Unlike the method of moments, the moment-based approach does not require pooled time-series and cross-section data for its implementation. The flexible moment-based approach also has the advantage over conventional econometric production models in that it does not impose arbitrary restrictions on the stochastic structure of production.

The moment-based approach can be understood at two levels. On the purely theoretical level, it can be established that if the range of a random variable such as output is finite, then the moments of output exist and uniquely determine its probability distribution function (see Rao, 1973:106). This sufficient condition for moments to determine a distribution holds for economic variables such as output, revenues, or profits. Thus all economically relevant characteristics of the technology are embodied in the relationships between inputs and the moments of the probability distribution of output, revenue, or profits, and the behavior of the firm under production uncertainty can be defined in terms of these relationships. On the practical level, the distribution of a random variable such as output can be approximated with a function of a small number of central moments. Such moment-based approximations to the stochastic technology are useful in applied research.

To demonstrate these points, consider the probability distribution $F[q|x,z,\phi]$, where q is output, given input vectors x and z and a parameter vector ϕ. The moments of the output distribution are defined as

$$\mu_1 = \mu_1^*(x,z,\phi) = \int q \, dF[q|x,z,\phi],$$

$$\mu_i = \mu_i^*(x,z,\phi) = \int (q - \mu_1^*)^i \, dF[q|x,z,\phi], \quad i \geq 1 \tag{4.1}$$

By the assumption that q is finite, it can be shown that the moments of its distribution exist and uniquely determine the distribution (Rao, 1973). Thus there exist moment functions

$$\mu_i = \mu_i(x,z,\beta_i), \quad i = 1,2,\ldots \tag{4.2}$$

such that the output distribution can be written

$$F[q|x,z,\phi] = F[q|\mu_1(x,z,\beta_1), \mu_2(x,z,\beta_2), \ldots] \tag{4.3}$$

It follows that the parameters of the moment functions, the β_i, imbed the properties of the output distribution and its parameters ϕ.

The general representation of the moments in equation (4.2) is flexible in the sense that each moment function depends on a distinct parameter vector. Thus across-moment restrictions are not imposed on the model, as they are in conventional production function models, and hypothesized restrictions can be tested using the model. It should be emphasized again that the purpose of this model is to obtain a flexible representation of the technology that does not impose unwarranted restrictions on the stochastic technology. For example, if it were known that the output distribution were lognormal, restrictions on the β_i across equations would be implied by the underlying parameters ϕ of the distribution. Moreover, if such restrictions were known, estimation efficiency could be increased by imposing them when the model is estimated.

To illustrate further, consider the stochastic generalization of the neoclassical production function, $q = q(x, z, \epsilon)$, where the error term ϵ is included to represent stochastic shocks to the production process. The distribution of ϵ is defined as $\Phi(\epsilon)$. The mean of output is then

$$\mu_1 = \int q(x, z, \epsilon)\, d\Phi(\epsilon),$$

and the variance of output is

$$\mu_2 = \int (q - \mu_1)^2\, d\Phi(\epsilon),$$

so the effect of a change in input x_k on the variance of output is

$$\begin{aligned} \frac{\partial \mu_2}{\partial x_k} &= 2\int (q - \mu_1)\left(\frac{\partial q}{\partial x_k} - \frac{\partial \mu_1}{\partial x_k}\right) d\Phi(\epsilon) \\ &= 2\,\mathrm{cov}(q, \partial q/\partial x_k), \end{aligned}$$

where $\mathrm{cov}(a, b)$ is the covariance of a and b. (In the above derivation it is assumed that the function q is twice continuously differentiable, so as to allow differentiation under the integration sign.) It follows that $\partial\mu_2/\partial x_k$ depends on the covariance between output and the marginal product of x_k.

In many econometric models, the covariance in the equation just stated is predetermined by the model's structure. Consider, for example, the multiplicative error model

$$q = m(x)\, e^u, \tag{4.4}$$

where $m(x)$ is interpreted as a neoclassical production function and u is a random error term. If u were a normal variate, (4.4) would be the lognormal model frequently used in econometric production studies. Letting $E[\cdot]$ denote the expectation operator, the mean of output is

$$\mu_1 = E[q] = m(x)\, E[e^u],$$

the variance is

$$\mu_2 = E[q - \mu_1]^2 = m(x)^2 E[(e^u - E[e^u])^2],$$

and in general the ith moment about μ_1 is

$$\mu_i = m(x)^i E[(e^u - E[e^u])^i]$$

Thus the multiplicative error model implies that the mean and the higher moments of the probability distribution of output are functions of inputs through the function $m(x)$. Since $\partial q/\partial x_k = \partial m/\partial x_k e^u$,

$$\frac{\partial \mu_2}{\partial x_k} = 2 \text{ covariance}(m, \partial m/\partial x_k)$$

$$= 2m(x)\frac{\partial m}{\partial x_k} E[(e^u - E[e^u])^2] > 0$$

More generally, this model imposes restrictions on the derivatives of all higher moments of output. The set of restrictions, or maintained hypotheses, can be expressed conveniently in terms of the elasticities of the moment functions with respect to inputs. From the previous equation it follows that, for $\mu_i \neq 0$, the elasticity of the ith moment with respect to x_k is

$$\nu_{ik} \equiv \frac{\partial \mu_i}{\partial x_k}\frac{x_k}{\mu_i} = i\frac{\partial m}{\partial x_k}\frac{x_k}{m} = i\nu_{1k}, \quad i \geq 2$$

Thus in the multiplicative error model the elasticity of the ith moment with respect to the kth input is proportional to the mean production elasticity of x_k. These restrictions are important in the analysis of stochastic technology because they arbitrarily constrain the behavior of the firm.

The flexible moment-based approach relaxes all such restrictions on the moment functions. The cost of flexibility is an incidental parameter problem, in the sense that there are as many parameter vectors as moments. An empirically useful representation of a stochastic technology cannot be based on a very large number of parameters. This problem can be resolved by applying the principle that empirical research should strive to obtain a good approximation to the true underlying relationships identified by theory. While there is no general means of determining what a "good enough" approximation is, evidence from several sources suggests that in approximating distributions, the first three moments are likely to give a good approximation to the extent that the basic shape characteristics of a unimodal distribution—location, dispersion, and asymmetry—are represented. Kendall and Stuart (1977) show that a probability distribution can be approximated to the nth degree by an nth-degree polynomial whose coefficients depend on the first n moments of the distribution. Thus there is a justification for using the first few moments of an empirical distribution to approximate the true unknown distribution. Kendall and Stuart conclude that "approximations of this kind often turn out to be remarkably good, even when only the first three or four moments are equated" (p. 90). The fact that many unimodal distributions can be adequately represented as functions of four or fewer moments is also demonstrated by the Pearson system of distributions (Kendall and Stuart, 1977, chap. 6). Members of the Pearson system are known to be functions of not more than the first four moments. Another argument in favor of moment-based approximations for the analysis of firm behavior has been put forward by Anderson, Dillon, and Hardaker (1977:97–98), who observe that when expected utility is approximated by a Taylor series (an approach discussed later in this chapter), going beyond a third- or fourth-degree approximation usually adds little to the accuracy of the approximation.

The theoretical moment functions are translated into an econometric model by observing that a random variable can be written as its mean plus a random variable with mean zero. Thus

$$q_j = \mu_1(x_j, z_j, \beta_1) + e_{1j}, \quad E[e_{1j}] = 0, \tag{4.5}$$

and similarly

$$(e_{1j})^i = \mu_i(x_j, z_j, \beta_i) + e_{2j}, \quad E[e_{2j}] = 0, \tag{4.6}$$

where $j = 1, \ldots, N$ indexes the sample observation. It should be noted that the effects of random events in the production process that are not measured by x_j and z_j, such as weather and pest damage, are represented in (4.5) and (4.6) by the error terms e_{ij}. The data can come from a cross section of observations, a time series, or pooled data. The type of data determines appropriate covariance assumptions. An important assumption underlying econometric estimation of the model is that all firms in the sample face a similar stochastic technology, in the sense that any systematic technological differences can be accounted for by using observable explanatory variables.

A heuristic interpretation of the principles underlying the estimation procedures is presented below. The statistical assumptions that must be made to establish the statistical properties of estimates of the β_i of the model in equations (4.5) and (4.6), as well as the proof of the limiting properties of the estimator to be discussed here, are examined in detail in Antle (1983b). For an in-depth study of the various statistical assumptions that can be made for the consistency and asymptotic normality of heteroscedastic regression models such as this one, see White (1984).

Consider a least squares regression of q_j on x_j and z_j. The resulting estimate of β_1 is consistent, implying that the regression residuals are consistent. By Slutsky's theorem, it follows that a residual taken to any power is a consistent estimate of $(e_j)^i$, that is, $\text{plim}(\tilde{e}_j)^i = (e_j)^i$. It follows that by replacing e_j in (4.6) with the residual $\tilde{e}_j$, the corresponding regression yields a consistent estimate of β_i. It is straightforward to show, however, that the error terms in equations (4.5) and (4.6) are heteroscedastic. By constructing a consistent estimator for the heteroscedastic variances of the error terms, a feasible generalized-least-squares (GLS) estimator can be utilized which is consistent and asymptotically normally distributed. This heteroscedasticity can be accounted for in several ways, one of which is to utilize White's (1984) heteroscedastic-consistent covariance estimation method. A more efficient, but comparatively more costly, procedure is to account for the exact heteroscedastic structure of this particular model—a procedure outlined below, and discussed in Antle (1983b).

In implementing the model, a specific functional form must be chosen for the moment functions. In this study the moment functions are specified as quadratic in the variables and linear in the parameters; this function is attractive in that its derivatives are linear in the variables and parameters. Letting X_j denote the vector containing the variables in the quadratic expansion, the estimation algorithm to obtain consistent, asymptotically normal parameter estimates proceeds as follows:

(A) Compute the regression

$$q_j = X_j\beta_1 + e_{1j}$$

to obtain consistent estimates of the residual $\tilde{e}_{1j}$.

(B) Compute the regressions

$$(\tilde{e}_{1j})^i = X_j\beta_i + e_{ij}, \quad i = 2,3$$

to obtain consistent estimates of the residuals $\tilde{e}_{ij}$, $i = 2,3$.

(C) Define heteroscedastic variances of the e_{ij}, $i = 1,2,3$, as the functions $X_j\gamma_i$, $i = 1,2,3$. To obtain consistent estimates of these variances, estimate the regressions

$$(\tilde{e}_{ij})^2 = X_j\gamma_i + v_{ij}, \quad E[v_{ij}] = 0,$$

subject to $X_j\gamma_i > 0$. The latter condition forces the estimated variances $X_j\tilde{\gamma}_i$ to be positive.

(D) Repeat steps (A) and (B) with the data weighted by $(X_j\tilde{\gamma}_i)^{.5}$.

In computational efficiency and accuracy step (C) has been found to be an improvement over the corresponding step 2 in Antle (1983b). To use White's heteroscedastic-consistent estimator, the White covariance estimation method would replace the one generated in steps (B) and (C).

The above estimation algorithm requires specialized computer programs. The empirical results reported below are based on programs written by the author in the programming language SPEAKEASY, and run on a VAX-750 with the VMS operating system. Step (C) of the algorithm was implemented through use of the MINOS program (see Murtaugh and Saunders, 1977). These procedures may also be implemented by using standard econometric software, such as SAS or SHAZAM, in conjunction with a nonlinear optimization program that can impose linear inequality constraints. White's estimator is available in SHAZAM.

Accounting for Sequential Decisions

It was shown in chapter 3 that input decisions associated with agricultural production processes are likely to be made sequentially because of the uncertainty in the production process and the time dimension associated with biological processes. Consequently, a dynamic econometric production model defined in terms of a sequence of stochastic outputs and input decisions has a triangular structure: output in a given period depends on current inputs, past inputs, and past outputs; inputs in that period depend on current input prices and on the information available at that time, including past inputs, outputs, prices, and the state of the crop. As a result of this structure inputs may be correlated with outputs, in violation of the assumption made earlier that inputs are exogenous to output. Whether or not this problem arises in practice depends on the assumptions made about the way decision makers use information, and on the data the econometrician has available. Antle (1983a) showed that a simultaneous equation estimator is *not* required when decisions are made sequentially if (1) decision makers *do not* update their information about the crop state over time, or (2) output and input data are available to the econometrician from each stage in the sequential decision process.

To clarify the issues involved, it is useful to consider a simple example in which inputs other than pesticides are exogenous to output. Let the production process be made up of T stages, with corresponding inputs x_t, random shocks ϵ_t, and outputs q_t. The output of each stage is distributed according to $F_t[q|x_t, q_{t-1}]$. Final output is q_T and inputs x_t, $t = 1, \ldots, T$ are observed by the econometrician, but intermediate outputs q_t, $t < T$, are not observed. The econometrician's goal, given inputs, is to characterize the distribution of final output. The joint distribution of all outputs is $F_T[q_1, \ldots, q_T|x_1, \ldots, x_T]$. Integrating over outputs the mean of final output is

$$\int \ldots \int q_T \, dF_T[q_1, \ldots, q_T|x_1, \ldots, x_T] = \mu_1[x_1, \ldots, x_T],$$

so

$$q_T = \mu_1[x_1, \ldots, x_T] + e_1, \quad E[e_1] = 0$$

Note that if random events such as pest infestations are not observed by the econometrician, they are embedded in the error term e_1.

Consider the use of pesticides purely as an "insurance" input—that is, as the case in which the pest management strategy is based on prior expectations about the pest population, without information updating during the season. In this case condition (1) is satisfied, and pest management input is a function of information Ω_0 which is available before production begins. Assuming the decision maker chooses inputs to maximize an objective function such as expected utility of profit, the firm's input demand functions take the form

$$x_t = x_t[p_T, w_1, \ldots, w_t, \Omega_0] \tag{4.7}$$

Hence the x_t are not functions of intermediate outputs and are not correlated with e_1. It follows that single-equation estimates of the output distribution, using the methodology described in the previous section, are consistent.

Suppose now that growers use an IPM program which specifies decision rules for pesticide application based on statistically reliable field samples of pest populations. Pest management decisions are then functions of the crop state as it evolves over the growing season, and the input demand functions take the form

$$x_t = x_t[p_T, w_1, \ldots, w_t, \Omega_0, q_1, \ldots, q_{t-1}] \tag{4.8}$$

Since q_T and q_t, $t < T$, generally are correlated, it follows that inputs and final output are correlated and therefore single-equation estimates of the output distribution are inconsistent. In order to obtain consistent estimates of the output distribution, the correlation between inputs and final output must be taken into account.

Thus when farmers follow an integrated pest management program, or more generally when farmers make decisions sequentially, intermediate inputs such as pesticides are endogenous. There are two classes of estimators that can be used to account for input endogeneity in dynamic production models. One class exploits the triangular structure of dynamic models, as discussed by Antle and Hatchett (1986). A computationally more convenient but less efficient estimator can be constructed using the nonlinear two-stage least squares estimator, or N2SLS, developed by Kelejian. N2SLS is appropriate here because the model is based on quadratic representations of the moment functions which are *nonlinear in the variables*, but *linear in the parameters*. To utilize the N2SLS estimator, one simply modifies the estimation algorithm described above. The first step in the modification is to perform the first-stage regressions of the N2SLS estimation procedure by regressing each right-hand-side endogenous variable on the exogenous and predetermined variables in the system. The above estimation procedure is then followed, with the fitted values from the first-stage regressions replacing the endogenous variables on the right-hand side of the moment equations.

Measuring Risk Attitudes

The evaluation of productivity and producer welfare under uncertainty requires knowledge of producers' objective functions. Although little is known about these functions, a considerable amount of research has been devoted to the modeling of decision making under uncertainty. As discussed in chapter 3, within the expected utility approach the objective function can be categorized in terms of the decision maker's risk attitudes. Here an econometric methodology is developed for the estimation of producers' risk attitudes within the expected utility approach. Given empirical estimates of risk attitudes, the empirical analysis of productivity and welfare under uncertainty can be pursued.

The econometric methodology developed below is designed to produce estimates of the *distribution* of risk attitudes in a population of producers who utilize a similar production technology. This approach differs from other attempts to quantify producers' risk attitudes by focusing on the characteristics of a population of producers rather than on the individuals in the population (see Hazell, 1982; Pope, 1982; Binswanger, 1982; and Newbery and Stiglitz, 1981 for discussions of the literature). There are several motivations for the present approach. The first—and pragmatic—is that unless producers' risk attitudes are assumed to be identical, it would not be possible to estimate each individual's risk attitude coefficients unless a sufficiently long time series were available for each individual. If each individual's utility function contains K parameters, and there are N individuals in the sample, at least K observations per individual would be required—many more if accurate estimates were to be obtained. By estimating the parameters of the distribution of risk attitudes, accurate estimation is possible as long as the number of risk attitude parameters is substantially less than $N \cdot K$. A second important motivation for this approach concerns the usefulness of risk attitude estimates for policy analysis. Generally the population's characteristics, and not individual risk attitudes, are relevant to policy analysis. For example, in the analysis of the adoption of an integrated pest management technology, the focus typically is on the "representative" farmer in the population, rather than on any individual.

A third motivation is to provide an econometric alternative to methods based on hypothetical questions and experiments. These non-econometric methods have the advantage of not requiring the assumptions that must be made to identify and estimate a structural econometric model of producer behavior. However, risk attitude estimates that are not based on observed economic behavior also have certain limitations. In particular, the risk characteristics of hypothetical or experimental decisions do not necessarily correspond to the actual production decisions faced by farmers. It is not known how relevant to the analysis of producer behavior or to policy analysis are the findings from methods which abstract from farmers' actual production decisions. Moreover, experiments which offer subjects realistic gains and losses may be prohibitively costly when conducted where real incomes of producers are high. Such experiments with large payoffs also may raise ethical problems for researchers; how does the researcher, for example, decide who gets the chance to win a large sum of money?

While an econometric approach based on production survey data is nonexperimental and requires those assumptions that necessarily underlie a structural econometric model, it has certain advantages over experimental methods. Risk attitude estimates based on observations of farmers' actual production decisions do correspond to the kind and degree of risk farmers actually face, and therefore are relevant to analysis of producer behavior and policy. The availability of an econometric methodology based on production survey data also allows risk attitudes to be estimated when it is too costly, or otherwise infeasible, to conduct experiments. An econometric approach permits researchers to draw upon existing statistical theory and hypothesis-testing procedures in the analysis of risk attitudes. Hypotheses implied by theory, such as the hypothesis of decreasing absolute risk aversion, could be formally tested.

The General Model

The econometric methods developed here require firm-specific production data which measure output and input quantities, their prices, and other observable technological characteristics of the firms. To simplify the presentation, the production data are assumed to be generated by farmers solving a static maximization problem: inputs are chosen conditional on information available before production begins; and input decisions are not made sequentially during the season. The analysis can be generalized to the case in which input decisions are made sequentially by utilizing the N2SLS procedure for the estimation of the technology parameters.

The stochastic technology can be represented by a joint probability distribution of output, revenue, or profit; the most general model based on profit is the one discussed here. Define q_j as a vector of outputs of farm j, p_j as the conformable output price vector, x_j and w_j as variable input quantity and price vectors, and z_j as a fixed input vector. Profit is defined as the returns to fixed factors: $\pi_j = p_j q_j - w_j x_j$. The distribution of profit is determined by the joint distribution of outputs and prices conditional on variable input choices and fixed inputs, and is written

$$F(\pi | x_j, z_j, \phi), \quad j = 1, \ldots, N, \tag{4.9}$$

where ϕ represents the parameters of the profit distribution.

In the models discussed in this study, the same profit (or revenue) distribution function and the same parameter vector α applies to each farmer in the population because all farmers are assumed to produce with the same stochastic technology, to face the same price distributions, and to form expectations based on those price and output distributions. However, as Schultz (1975) has emphasized, farmers may not be in this kind of a rational expectations equilibrium. When events such as rapid technological change occur, farmers will acquire information at different rates as a function of their human capital endowments. Thus all farmers will not have the same subjective expectations. The existing evidence on this question suggests that the assumption of rational expectations depends on the degree of equilibrium or disequilibrium experienced by farmers. Grisley and Kellogg

(1983) found evidence that farmers' subjective expectations were accurate estimates of objective distributions in a state of relative economic equilibrium, whereas Pingali and Carlson (1985) found that human capital plays an important role in the accuracy of subjective expectations in the presence of technological change. It follows that the assumption that all farmers face the same profit distribution may have to be modified in cases where the population under investigation is experiencing an information disequilibrium.

In the moment-based model risk attitudes are defined in terms of the derivatives of the utility function. The jth farmer's utility function is

$$U_j = U(\pi_j, \gamma_j), \tag{4.10}$$

where γ_j is a vector of parameters representing the individual's risk attitudes. Generalizing the results of Pratt (1964), γ_j can be interpreted as measuring Arrow-Pratt and downside risk aversion:

$$U(\pi_j, \gamma_j) = e^c \int \exp\left[\int \left\{ \int (DS - AP^2) \right\} \right],$$

where c is a constant of integration; $AP = -U^2/U^1$, where U^i is the ith derivative of U with respect to π_j; and $DS = U^3/U^1$. AP measures Arrow-Pratt absolute risk aversion, and DS can be interpreted as a measure of downside risk aversion (see Menezes, Geiss, and Tressler, 1980).

The utility function is often expressed in terms of wealth (that is, changes in wealth in the current period) rather than income or profit. Note, however, that if a utility function of wealth $V(W + \pi)$ is defined, that

$$\frac{\partial V(W + \pi)}{\partial \pi} = \frac{\partial U(\pi)}{\partial \pi}$$

Thus, for the analysis of current-period input decisions which affect π and not W, the two types of objective functions yield the same implications. Since the decision maker's risk attitudes are interpreted in terms of the derivatives of the utility function, and are measured as such, it is impossible to differentiate these two specifications of the objective function without data on the farmer's wealth. Wealth data typically are not available in production studies.

Expected utility is, from equations (4.9) and (4.10),

$$EU_j = \int U(\pi, \gamma_j)\, dF(\pi | x_j, z_j, \phi) = u[x_j, z_j, \gamma_j, \phi] \tag{4.11}$$

Assuming the objective function is globally concave in x_j and has a unique interior solution, the first-order conditions for maximization of the value function give (implicitly) the optimal variable input quantities:

$$\partial u[x_j, z_j, \gamma_j, \phi]/\partial x_j = \delta[x_j, z_j, \gamma_j, \phi] = 0 \tag{4.12}$$

Equations (4.9) through (4.12) provide the basis for deducing the statistical properties of cross-sectional production data. The structural model is the profit distribution (4.9) and the first-order condition (4.12). The endogenous variables are the variable inputs plus profit. The exogenous variables are the fixed inputs z_j, the utility parameters γ_j, and the technology and price distribution parameters ϕ. Generally, the z_j are observable, while the γ_j and ϕ vectors are not observed (prices represented by ϕ can be observed, but parameters of price distributions generally cannot be).

To begin the analysis of econometric estimation, consider first the case of identical risk attitudes in the producer population—that is, $\gamma_j = \gamma$ for all j. In this case, the econometric problem is to estimate the parameter vectors γ and ϕ. Under this assumption, the model given by (4.9) and (4.12) is a generalization of the mean-variance model discussed by Just and Pope (1979a). Equations (4.9) and (4.12) can then be interpreted as a (block) triangular system with dependent variables π_j and x_j. Given a parametric specification of the profit distribution or of its moment functions and sufficient covariance restrictions, the technology parameters ϕ are necessarily identified (see Hsiao, 1983 for a discussion of identification of such systems). However, identification of the factor demand equation parameters, and by implication the utility parameters γ, is not necessarily possible. For example, if the simultaneous block of factor demand equations were linear in the parameters, there were more than one variable input, and there were no other identifying restrictions, the factor demand equations would be underidentified. By utilizing the across-equation restrictions inherent in the model, the factor demand equations may be identified. If the demand equations are nonlinear in the parameters, the nonlinearities also may aid in identification.

In the more realistic case of heterogeneous risk attitudes ($\gamma_i \neq \gamma_j$ for $i \neq j$), it is impossible to estimate each individual's vector γ_j unless a sufficiently long time series is available for each individual. However, it may be possible to estimate the parameters of the *distribution* of risk attitudes in the producer population with fewer data. In this case the input vector x_j is a random variable because it depends on γ_j (see 4.12). Similarly, the firm's vector of fixed inputs can be viewed as distributed in the population of producers. Define the joint distribution of x_j, z_j, and γ_j in the population as

$$g(x, z, \gamma | \bar{x}, \bar{z}, \theta), \tag{4.13}$$

where $\bar{x}$ and $\bar{z}$ are the mean input vectors in the population and θ is a parameter vector. Using (4.12) and (4.13), the population mean of the implicit factor demand equations is

$$\iiint \delta[x, z, \gamma, \phi] \, g(x, z, \gamma | \bar{x}, \bar{z}, \theta) \, dx \, dz \, d\gamma \equiv \bar{\delta}[\bar{x}, \bar{z}, \theta, \phi] = 0 \tag{4.14}$$

Replacing $\bar{x}$ and $\bar{z}$ in (4.14) with their observed values, define

$$\bar{\delta}^*[x_j, z_j, \bar{x}, \bar{z}, \theta, \phi] \equiv \bar{\delta}[x_j, z_j, \theta, \phi] - \bar{\delta}[\bar{x}, \bar{z}, \theta, \phi]$$

Note that $\bar{\delta}^*$ measures the difference between $\bar{\delta}[x_j, z_j, \theta, \phi]$ and the mean first-order condition $\bar{\delta}[\bar{x}, \bar{z}, \theta, \phi]$, and that generally $E[\bar{\delta}^*] \neq 0$. In addition, define

$$\bar{\bar{\delta}}[x_j, z_j, \bar{x}, \bar{z}, \theta, \phi] \equiv E\{\bar{\delta}[x_j, z_j, \theta, \phi]\} - \bar{\delta}^*[x_j, z_j, \bar{x}, \bar{z}, \theta, \phi] = \epsilon_j, \tag{4.15}$$

where $E(\epsilon_j) = 0$. Thus $\bar{\bar{\delta}}$ represents the difference between the expectation of $\bar{\delta}_j$ and the term $\bar{\delta}^*$ defined above. Using the definition of $\bar{\delta}^*$,

$$E\{\bar{\bar{\delta}}[x_j, z_j, \bar{x}, \bar{z}, \theta, \phi]\} = \bar{\delta}[\bar{x}, \bar{z}, \theta, \phi] = 0$$

These relationships show that the firm-specific behavioral equation (4.12), which cannot be estimated in a single cross section, can be transformed into a behavioral equation (4.15) defined in terms of the population parameters θ and ϕ that can be estimated with a single cross section of data. The econometric problem then is to identify and estimate the parameters of (4.15). The structure of the production model composed of equations (4.9) and (4.15) is similar to the structure of the model (4.9) and (4.12) with homogeneous risk attitudes, and the identification problem is identical. The technology parameters are necessarily identified, but the vector θ can be identified only through across-equation restrictions, nonlinearities, or other restrictions.

Under appropriate distributional assumptions, the parameters of equations (4.9) and (4.15) can be estimated jointly using procedures appropriate for systems of implicit, nonlinear equations (Gallant and Jorgenson, 1979). However, such procedures generally involve costly iterative solution of nonlinear equation systems. An alternative estimation approach exploits the recursive structure of the model. First, a consistent estimate $\tilde{\phi}$ of the technology parameters ϕ is obtained, for example, using the procedures described in Antle (1983b). Second, an appropriate estimator is used to consistently estimate θ, given $\tilde{\phi}$. One such estimation approach is the Generalized Method of Moments (GMM) presented by Hansen (1982). The GMM approach is suitable for this problem because it allows the first-order conditions to be estimated in either implicit form, as in equation (3.12), or explicit form, and allows endogeneity of the inputs to be accounted for by using instrumental variables techniques. Hansen shows, under a set of general conditions on the data and the functions in the model, that an instrumental variable estimator generally exists and can be used to obtain consistent, asymptotically normal parameter estimates. A summary of the GMM estimator for this model is presented in Antle (1987, appendix 1).

The Moment-Based Model

A tractable empirical model must be based on approximations to the stochastic technology and the decision maker's objective function. In this section the moment-based approximation of the output distribution developed above is used to derive a model that can be estimated by using established econometric methods.

To motivate the approach, suppose the first-order conditions for expected utility maximization can be written

$$D_{1jk} = -D_{2jk}r_{2j} - \cdots - D_{mjk}r_{mj} + \epsilon_{0jk} \tag{4.16}$$

where

$$D_{ijk} = \partial\mu_{ij}/\partial x_{kj}, \quad E(\epsilon_{0jk}) = 0, \quad E(\epsilon_{0jk}, \epsilon_{0jk}) = \theta_{0kk'}$$

The r_{ij} represent the jth farmer's risk attitudes as functions of the derivatives of the utility function. The error term ϵ_{0jk} represents possible random errors in maximization. Suppose further that the distribution of the risk attitudes in the population can be written

$$r_{ij} = \theta_1^i + \epsilon_{ij}, \quad E[\epsilon_{ij}] = 0$$

Then, substituting this equation into (4.16),

$$D_{1jk} = -D_{2jk}\theta_1^i - \cdots - D_{mjk}\theta_1^m + \omega_{jk} \tag{4.17}$$

where

$$\omega_{jk} = \epsilon_{0jk} - D_{2jk}\epsilon_{2j} - \cdots - D_{mjk}\epsilon_{mj}, \quad \text{and } E[\omega_{jk}] = 0$$

In order to be able to estimate (4.17) as a regression, its statistical properties must be established. Observe that the D_{ijk} are functions of the inputs x_{jk}, which in turn are functions of the individual's risk attitudes. Thus the D_{ijk}, $i > 1$, are correlated with the ϵ_{ij}. Consistent estimation of (4.16) must account for this correlation. Moreover, it is clear that ω_{ik} is heteroscedastic. Estimation also must account for the structure implied by the covariances of the ϵ_{ij}.

We proceed now to formally derive the above model for risk attitude estimation. Using equation (4.3), expected utility can be expressed as

$$\begin{aligned} EU_j &= \int U(\pi_j, \gamma_j)\, dF[\pi | \mu_1(x_j, z_j, \beta_1), \mu_2(x_j, z_j, \beta_2), \ldots] \\ &= u[\mu_1(x_j, z_j, \beta_1), \mu_2(x_j, z_j, \beta_2), \ldots, \gamma_j] \end{aligned} \tag{4.18}$$

Under the assumption that output is finite, the output distribution can be approximated in terms of the first m moments. Let

$$\mu_j^m = (\mu_1(x_j, z_j, \beta_1), \mu_2(x_j, z_j, \beta_2), \ldots, \mu_m(x_j, z_j, \beta_m)) \tag{4.19}$$

The first-order condition for expected utility maximization is then

$$\sum_{i=1}^{m} (\partial u[\mu_j^m, \gamma_j]/\partial \mu_{ij})(\partial \mu_i[x_j, z_j, \mu_i]/\partial x_{jk}) = 0, \quad k = 1, \ldots, n, \tag{4.20}$$

where x_{jk} is the kth element of x_j. Thus, using the moment-based approximation to the profit distribution, expected utility can be written in terms of the derivatives of the expected utility function and the derivatives of the moment functions.

The first-order condition (4.20) provides the basis for the estimation of risk attitudes. Letting

$$r_{ij} = (\partial u[\mu_j^m, \gamma_j]/\partial \mu_{ij})/(\partial u[\mu_j^m, \gamma_j]/\partial \mu_{1j}), \quad i > 1, \tag{4.21}$$

$$D_{ijk} = \partial\mu_i[x_j, z_j, \beta_i]/\partial x_{jk}, \quad i = 1,2,\ldots \tag{4.22}$$

equation (4.20) can be written as equation (4.16) is.

Following the GMM approach to estimation described above, the estimation strategy is to obtain consistent estimates $\tilde{\beta}_i$ of the β_i and then estimate the behavioral equations (4.16) given the $\tilde{\beta}_i$. The D_{ijk} are observed quantities (given the $\tilde{\beta}_i$), but the r_{ij} are unobserved random variables because they depend on x_j, z_j, and γ_j. Define the distribution of r_{ij} in the population as follows:

$$r_{ij} = \theta_1^i + \epsilon_{ij}, \quad E[\epsilon_{ij}] = 0, \; E[\epsilon_{ij}\epsilon_{gk}] = \theta_{igjk} \tag{4.23}$$

The r_{ij} represent the ith dimension of risk attitudes in terms of the expected utility function. For $i = 2,3$, there are two characteristics of risk attitudes represented by the r_{ij}; it is shown below that these characteristics can be interpreted in terms of Arrow-Pratt risk aversion and downside risk aversion. The parameter θ_1^i measures the mean of r_{ij} and thus represents the mean risk attitude in the population in terms of the ith characteristic. The variance and covariance parameters also provide important information about behavior.

The assumption that risk attitudes are distributed according to equation (4.23) requires some clarification. Note that, in general, an *individual's* risk attitudes are independent of wealth or income if and only if the individual's preferences exhibit constant absolute risk aversion (CARA). Under CARA, variation in preferences could be a result of random intrapersonal or interpersonal variation, but not of changes in wealth. If CARA does not hold, however, variation across individuals could be a result of differences in wealth as well as differences in the underlying utility parameters. In making assumptions such as those embodied in (4.23), therefore, one must interpret the parameters of the distribution of the r_{ij} as reflecting the underlying factors that determine the distribution of risk attitudes in the population, which can include variations in wealth or other socioeconomic factors affecting risk attitudes, as well as randomness in the utility parameters themselves.

Equation (4.23) is an important part of the model because it defines the properties of the distribution of farmers' risk attitudes across individuals and across risk-attitude characteristics. Note that

$\theta_{iijj} \equiv$ variance of r_{ij} for individual j,

$\theta_{igjj} \equiv$ covariance of r_{ij} and r_{gj} for individual j,

$\theta_{iijk} \equiv$ covariance of r_{ij} and r_{ik} between individuals j and k,

$\theta_{igjk} \equiv$ covariance of r_{ij} and r_{gk} between individuals j and k.

Thus these elements of the covariance matrix provide information about the nature of the randomness in behavior. This information can be used to test hypotheses about the nature of behavioral variation. It has been hypothesized that individuals' preferences are random; this concept has been the basis for theories of random utility (see McFadden, 1983 and citations therein). In other words, variation in the population could result from intrapersonal variation. It has also been hypothesized that each individual has stable preferences over time, but that preferences vary randomly among individuals in the population; thus variation in the population can also be hypothesized to be a result of interpersonal differences in preferences. These alternative hypotheses suggest a variety of different models:

Model B. All individuals have fixed preferences, randomly distributed in the population:

$\theta_{igjj} = 0$, for all i and g,

$\theta_{igjk} \neq 0$, for all i and g, and for $j \neq k$.

Model C. All individuals have independent, but not identically distributed, random preferences:

$\theta_{igjk} \neq 0$, for all i and g, and for $j = k$,

$\theta_{igjk} = 0$, for all i and g, and for $j \neq k$.

Model D. Individuals have dependent, not identically distributed random preferences:

$\theta_{igjk} \neq 0$, for all i, g, j, k.

Model A implies that all variation in the population is a result of independent, intrapersonal variation, whereas Model B implies the opposite extreme, that all differences are interpersonal and correlated across individuals. Model C implies both intrapersonal and interpersonal differences, but preferences are independent; Model D relaxes the independence assumption. Note that in order to differentiate these alternative hypotheses, it is necessary to have sufficient data on a cross section of individuals over time so that both interpersonal and intrapersonal covariance can be estimated.

To simplify the remainder of this discussion, and to facilitate implementation of the estimated procedures, the assumption that risk attitudes are independent across individuals will be maintained. This assumption implies either models A or C; models B and D are ruled out, a priori.

Proceeding now with the econometric formulation of the model, rewrite equation (4.17) as

$$D_{1k} + D_{2k}\theta_1 = \omega_k, \quad k = 1, \ldots, n, \tag{4.24}$$

where D_{1k} is a vector of the D_{1kj}, D_{2k} is a matrix of the D_{ijk} $(i > 1)$, and ω_k is a vector of the ω_{jk}.

With the assumption that the r_{ij} are distributed in the population according to (4.23), the linear structure of equation (4.17) results in a random coefficient model. Given the technology parameter estimates, and given a sufficient set of exogenous and predetermined variables, the parameter vector θ_1 in (4.17) can be identified. From the theoretical model, the input choices x_{jk} are functions of γ_j, and therefore D_{2k} is correlated with the error term in (4.17). To obtain a consistent estimate of θ_1, then, an instrumental variables estimator is required. Let the vector y_j be a vector of instruments satisfying the condition that $E[\omega_{jk}\Omega y_j] = 0$, where Ω is the Kronecker product, and let Y be a matrix of the y_j. Then the instrumental variable estimator

$$\tilde{\theta}_1 = (Y'D_2)^{-1}Y'D_1 \tag{4.25}$$

provides a consistent estimate of the parameter vector θ_1 (here D_1 is a stacked vector of the D_{1k} and D_2 is a stacked matrix of the D_{2k}). The two-stage least squares version of the instrumental variables estimator, used in chapter 5, is obtained by defining $Y = Z(Z'Z)^{-1}Z'D_2$, where Z is a matrix of exogenous variables uncorrelated in the limit with risk attitudes, but correlated with D_2.

If we ignore the fact that the elements of D_2 were obtained from a first-stage estimation of the technology, the asymptotic covariance matrix for the instrumental variables estimator described above is

$$\sum = \text{plim}\, N(\Delta'Z'Z\Delta)^{-1}\Delta'Z'SZ\Delta(\Delta'Z'Z\Delta)^{-1}, \tag{4.26}$$

where $(Z'Z)^{-1}Z'D_2 = \Delta$ (Hansen, 1982). Given equation (4.23) and the independence of preferences across individuals, the elements of S in (4.26) are provided by

$$\begin{aligned} E[\omega_{jk}\omega_{j'k'}] &= \theta_{0kk'} + \sum_{i,g=1}^{m}\sum \theta_{ig}D_{ijk}D_{gjk'}, \quad j = j' \\ &= 0, \quad j \neq j' \end{aligned} \tag{4.27}$$

Consistent estimates of the elements of the covariance matrix can be obtained using a procedure similar to the one proposed by Hildreth and Houck (1968). For example, if the first-order conditions for two inputs x_{j1} and x_{j2} are represented in the model, the covariance matrix has the structure

$$S = \begin{bmatrix} S_{11} & S_{12} \\ S_{21} & S_{22} \end{bmatrix}$$

where the S_{ij} are diagonal and heteroscedastic. Consistent estimates of the elements of S_{11} and S_{22} can be obtained by computing instrumental variables estimates of the equation

$$\begin{aligned}(u_{jk})^2 &= \theta_{0kk} + \theta_{22}(D_{2jk})^2 + \theta_{33}(D_{3jk})^2 \\ &\quad + 2\theta_{23}D_{2jk}D_{3jk} + v_j\end{aligned} \tag{4.28}$$

where u_{jk} is the residual from the instrumental variable estimate of θ_1 and v_j is an error term. The fitted values from (4.28) are consistent estimates of the diagonal elements of S_{11} and S_{22}. Consistent estimates of the diagonal elements of S_{12} can be obtained by computing the regression

$$\begin{aligned}u_{j1}u_{j2} &= \theta_{012} + \theta_{22}D_{2j1}D_{2j2} + \theta_{33}D_{3j1}D_{3j2} \\ &\quad + \theta_{23}(D_{2j1}D_{3j2} + D_{2j2}D_{3j1}) + v_j\end{aligned} \tag{4.29}$$

Because the first-order conditions (4.16) are to be estimated by using the parameter estimates of the technology embedded in the D_{ijk}, the covariance matrix (4.26) is incorrect. Murphy and Topel (1985) have shown that, under the assumption that the random components in the first and second stages are uncorrelated, a straightforward adjustment of the covariance matrix is possible. Since in this model the errors in the first stage are attributable to the disturbances in the technology, they should indeed be independent of the errors in the second stage that are attributable to random variation in preferences. Parke (1986) has shown that efficient estimation of the first-stage parameters ensures this independence in any case. Details of the implementation of the Murphy-Topel procedure are presented in Antle (1987, appendix 1).

Interpretation of the Moment-Based Model

The preceding section shows that the moment-based approximation of the output distribution provides a means of obtaining consistent estimates of the distribution of the derivatives of the expected utility function. However, it is not immediately evident that those derivatives can be interpreted as measures of risk attitudes and thus how they are related to the parameter vector γ_j in the utility function (4.10). The present section will show how the r_{ij}, $i = 2, 3$, can be interpreted as measures of Arrow-Pratt and downside risk aversion. For this interpretation it will be assumed that the utility function (4.10) is analytic on a finite interval of the real line. Since output is finite, and since the units of measurement are arbitrary, any analytic function can be considered a candidate utility function. By scaling profit to lie in the interval on which the utility function is analytic, there is a convergent Taylor series expansion of the utility function for every value of π_j, and expected utility can be expressed as

$$EU_j = U(\mu_{1j}) + \sum_{i=2}^{\infty} U^i(\mu_{1j})\mu_{ij}/i! \tag{4.30}$$

where U^i is the ith derivative of U. It follows that

$$\partial EU_j/\partial\mu_i = U^i(\mu_{1j})/i!, \quad i > 1 \tag{4.31}$$

Furthermore,

$$\partial EU_j/\partial\mu_{1j} = U^1(\mu_{1j}) + \sum_{i=2}^{\infty} U^{i+1}(\mu_{1j})\mu_{ij}/i! = E[U_j^1] \tag{4.32}$$

Considering a third-order approximation, the expectation of marginal utility equals $U^1(\mu_{1j})$ when the marginal utility function is linear, that is, when $U_j^3 = 0$. Otherwise, expected marginal utility will over- (under-) estimate U_j^1 as U_j^3 is greater (less) than zero.

Recall from (4.22) that the r_{ij} measure the derivatives of the expected utility function with respect to the ith moment (given by 4.31), normalized by the derivative of expected utility with respect to the first moment (given by 4.32). Therefore,

$$r_{ij} = U^i(\mu_{1j})/E[U_j^1]i! \tag{4.33}$$

demonstrating that the signs of the derivatives of the *utility function* can be inferred from the derivatives of the *expected utility function* with respect to the profit moments. The r_{ij} are closely related to the Arrow-Pratt risk aversion measure $AP = -U^2/U^1$ and the downside risk aversion measure $DS = U^3/U^1$. Using $E[U^1]$ as a first-order approximation to $U^1(\mu_1)$, $-2\theta_1^2$ and $6\theta_1^3$ can be interpreted as the approximate mean Arrow-Pratt and downside risk attitudes in the population. The variances $2\theta_{22}$ and $36\theta_{33}$ indicate the degree of dispersion of risk attitudes in the population, and the covariance $-12\theta_{23}$ measures the degree of association between Arrow-Pratt and downside risk aversion.

The magnitude and range of risk attitudes can be interpreted in terms of the risk premium implied by the estimates of *AP* and *DS*. Extending the analysis of Newbery and Stiglitz (1981:73) to the case of a third-degree Taylor series approximation to the expected utility function, the risk premium ρ as a percentage of expected net returns (or the relative risk premium) is approximately

$$\rho/\mu_1 \simeq \mu_2 AP/2\mu_1 - \mu_3 DS/6\mu_1 \tag{4.34}$$

The degree to which a risk-averse farmer behaves differently from a risk-neutral farmer is determined by the degree to which *both* risk aversion *and* the risk characteristics of the technology combine to alter the risk-averse farmer's input choices. To illustrate, consider the first-order condition (4.19) for optimal input choice. For $m = 3$ this equation can be rewritten in percentage terms as

$$\nu_{1k} = AP\nu_{2k}\mu_2/\mu_1 - DS\nu_{3k}\mu_3/6\mu_1, \tag{4.35}$$

where

$$\nu_{ik} = (\partial\mu_i/\partial x_k)x_k/\mu_i$$

The left-hand side of (4.35) equals the expectation of the value of the marginal product minus the cost of factor x_k, as a percentage of net returns. The risk-neutral farmer equates ν_{1k} to zero to maximize expected net returns, but the risk-averse farmer takes into account the effects of inputs on the distribution of net returns, and thus generally does not equate ν_{1k} to zero.

As discussed in chapters 2 and 3, the amount that a risk-averse producer would have to be compensated to use the same amount of input as a risk-neutral farmer can be defined as the marginal risk premium. This premium must equal the difference between the expectation of the value of the marginal product and the factor price at the optimum, and therefore is measured by the right-hand side of (4.35) as a percentage of expected profit. If an input is risk-decreasing, in the sense of Pope and Kramer, the marginal risk premium is negative; it is positive if the input is risk-increasing. The marginal risk premium is relevant to policy analysis, as it indicates the magnitude of subsidy on a risk-increasing input that would be required to induce a risk-averse farmer to behave in a risk-neutral manner.

Equation (4.35) can be related to the discussion of input choice under uncertainty in chapter 2. Observe that (4.35) is the condition for EMU = 0. If the firm is risk-neutral, it equates ν_{1k} to zero (that is, EMU_0 in figure 2–5). With risk aversion and a risk-decreasing input, the expected marginal utility curve could be EMU_1, which would lie above the risk-neutral curve EMU_0. The marginal risk premium measures the vertical distance between these two curves. Alternatively, (4.35) can be related to figure 2–6. The left-hand side defines the relation of the EVMP curve to the factor price. Under risk neutrality they are equal, so $\nu_{1k} = 0$; under risk aversion, $\nu_{1k} \neq 0$ in equilibrium, and the distance between the EVMP and the SVMP curves is given by the left-hand side of (4.35).

Although Arrow-Pratt risk aversion has been used widely in the production literature, it is not necessarily the most relevant risk-aversion measure for the analysis of farm decision making under uncertainty. One limitation of the Arrow-Pratt concept is that it is valid only for the case in which a riskless alternative is available to the decision maker. Farmers' production decisions typically involve choices among many risky alternatives without consideration of a riskless alternative. This would be the case, for example, when the farmer is deciding how much pesticide to use. Given the decision to produce, the farmer faces production risk whatever amount of pesticide is chosen. Under these conditions Ross's (1981) stronger measure of risk aversion is relevant to production decisions, because it provides orderings of risky alternatives when a riskless alternative is not considered. Ross's conditions for strong risk aversion also can be defined in terms of the derivatives of the utility function. He shows that utility function U is strongly more risk-averse than V if and only if there exists a λ, for all π_1 and π_2, such that

$$U^2(\pi_1)/V^2(\pi_1) \geq \lambda \geq U^1(\pi_2)/V^1(\pi_2) \tag{4.36}$$

It should be noted that r_{2j}/r_{2k} can be interpreted as an approximation to the left-hand side of (4.36). Given estimates of risk attitudes, it is possible also to approximate marginal utility and thus to approximate the right-hand side of (4.36), and to check the above condition for strong risk aversion.

Ross also shows that the utility function exhibits strong decreasing absolute risk aversion when there exists a number λ such that

$$U^3(\pi)/U^2(\pi) \leq \lambda \leq U^2(\pi)/U^1(\pi),$$

with the inequalities reversed for increasing absolute risk aversion. Since this condition involves the ratios of utility function derivatives at the same value of π, it is possible to use the estimated distribution of r_{2j} and r_{3j} to check whether or not this condition is satisfied at the population mean and other points on the distribution.

Risk Attitude Estimation with Multiproduct Technologies

The model generating the data (as in equation [4.9]) was specified as multiproduct, by virtue of the random output vector q_j embedded in π_j. Generally, the methods described in this study are applicable to either single-product or multiproduct farms because the expected utility model has the same structure in both cases. However, in practice econometric analysis may be possible only with data for one or a few products, even though the production process is truly multiproduct. This is especially likely to be the case in agriculture, for several reasons. First, when there are many outputs there are likely to be many inputs allocated to each production process, so the total number of inputs is large and a degrees-of-freedom problem may arise. Second, there may be many highly colinear inputs in multiproduct analysis, so that unless inputs are aggregated, econometric analysis may be difficult. Third, it is rare for data to be available on all production activities.

These difficulties raise the question of whether or not risk attitudes can be measured by using data for a subset of production activities. Clearly, the answer hinges on the "jointness" properties of the stochastic technology. The first-order conditions (4.12) show that in the general case all technology parameters and data are present in each equation, so data on all production activities are required. But if some input demand equations do not depend on all production activities' technology parameters and data, it should be possible to measure risk attitudes by the methods described above for the moment-based model without data for all production activities.

Consider the case in which there are G production activities, but data are available only for activity 1. The first-order condition for the choice of x_{jk}^{g} (jth farm, kth input, gth output) is, from (4.20),

$$\sum_{i=1}^{m} \frac{\partial u[\mu_j^m, \gamma_j]}{\partial \mu_{ij}} \frac{\partial \mu_i(x_j, z_j, \mu_i, \beta)}{\partial x_{jk}^{g}} = 0, \qquad (4.37)$$

$$k = 1, \ldots, n, \quad g = 1, \ldots, G,$$

where the moments are defined in terms of profit summed over all products. If the effects of an input on the moments of the first production activity equal the effects that input has on the moments of total profit, that is if

$$\frac{\partial \mu_i^1(x_j^1, z_j^1, \beta_i^1)}{\partial x_{jk}^1} = \frac{\partial \mu_i(x_j, z_j, \beta_i)}{\partial x_{jk}^1}, \quad i = 1, \ldots, m, \tag{4.38}$$

it would be possible to express the population first-order condition in terms of data for activity 1, and thus to estimate the risk attitude parameters of the population (here μ_i^1 is the ith moment of profit from product 1, μ_i is the ith moment of total profit, and other variables are defined accordingly).

It is shown in Antle (1987, appendix 3) that a necessary and sufficient condition for (4.38) to hold is that the moments of π^1 depend only on x^1 and z^1, and that the moments of $\pi^* = \pi^2 + \cdots + \pi^G$ and the covariances and higher product moments of π^1 and π^* do not depend on x^1 and z^1. This condition is defined as *exact stochastic nonjointness in inputs*, and does not require that π^1 and π^* be statistically independent. Unfortunately, the conditions for exact stochastic nonjointness are difficult to relate to the technological characteristics that one can observe in an actual farming system. There is an intuitively plausible and more easily interpreted approximate sufficient (but not necessary) condition for (4.38), however. Defined as *approximate stochastic nonjointness in inputs*, this condition holds if and only if

$$\begin{aligned} &F(\pi^1, \ldots, \pi^G | x_j, z_j, \alpha) \\ &\quad = F^1(\pi^1 | x_j^1, z_j^1, \alpha_1) F^*(\pi^2, \ldots, \pi^G | x_j^*, z_j^*, \alpha^*), \end{aligned} \tag{4.39}$$

where F denotes the joint probability distribution for profit from all G production activities, and F^1 and F^* are similarly defined. In Antle (1987, appendix 3) it is shown that (4.39) is approximately sufficient for (4.38) in the sense that it is exactly sufficient for distributions that can be defined in terms of their first three moments.

Equation (4.39) states that profit from product 1, π^1, is statistically independent of $\pi^2, \ldots, \pi^G$, and that the distribution of π^1 depends only on x_j^1 and z_j^1. These conditions are readily interpretable in terms of a farming system. For example, product 1 could be the farmer's main cash crop, and products $2, \ldots, G$ could be household production activities.

The Measurement of Efficiency, Welfare, and Technological Change

Once the technology and the distribution of risk attitudes have been estimated, the efficiency indexes discussed in chapter 3 can be used to analyze welfare and technological change. This section outlines a tractable empirical framework for the analysis of efficiency and welfare under uncertainty. The framework to be described is attractive because the first-order conditions are linear, and it therefore provides an explicit solution to the firm's optimization problem. The empirical model can be used to explore a variety of questions related to efficiency under uncertainty, including the effects of different degrees of risk aversion on efficiency and optimal input demands; how efficiency compares among groups of farms, such as those using or not using an IPM technology; and the effects of characteristics of the farm manager, like schooling, on efficiency. To illustrate the approach, the relative efficiency of two groups of farms is considered. Farms within each group are assumed to produce at the same efficiency, but efficiency may differ across groups.

Efficiency comparisons require that two basic components of the model be specified: the technology as embodied in the distribution of net returns, and the objective function $J(x, z, \alpha)$. In the illustration here, prices are treated as nonstochastic so that the output distribution can be used to characterize the technology. The objective function is specified as the maximization of expected utility, and the utility function is based on a third-order Taylor series approximation to the negative exponential utility function. The technical and allocative efficiency of each farm group can then be evaluated for specified risk attitudes.

It will be convenient to define the model in terms of profit normalized by the output price. Thus, define normalized profit as $\pi^* = q - w^*x$, where $w^* = w/p$. For nonstochastic prices, the moments of profit are then the moments of output (except for mean profit, which is adjusted by cost). The first three moments of output for the two farm groups (a) and (b) are defined as functions of inputs as follows:

$$\mu_1^j(x, z) = \int q \, dF^j(q | x, z, \phi), \quad j = a, b \tag{4.40}$$

$$\mu_i^j(x, z) = \int (q - \mu_1^j)^i \, dF^j(q | x, z, \phi), \quad i = 2, 3 \tag{4.41}$$

The objective function is specified to be the following Taylor series approximation to the negative exponential utility function, defined as a function of normalized profit:

$$U(\pi^*) = 1 - \exp(-\gamma(\mu_1 - w^*x)) - \exp(-\gamma(\mu_1 - w^*x)) \sum_{i=1}^{3} (-\gamma)^i (q - \mu_1)^i / i! \qquad (4.42)$$

The parameter γ is the Arrow-Pratt measure of absolute risk aversion in the utility function $U(\pi^*) = 1 - \exp(-\gamma\pi^*)$ from which (4.42) was derived. This utility function has a convergent Taylor series for all values of profit, and has its origin at zero, thus making it attractive for the computation of efficiency indexes.

Taking the mathematical expectation of (4.42) gives expected utility:

$$E[U(\pi^*)] = 1 - \exp(-\gamma(\mu_1 - w^*x)) - \exp(-\gamma(\mu_1 - w^*x))(\gamma^2\mu_2/2 - \gamma^3\mu_3/3) \qquad (4.43)$$

The first-order conditions for expected utility maximization imply that the following conditions must hold for the variable inputs:

$$\frac{\partial \mu_1}{\partial x_k} - w_k = \delta^{-1}\left(\frac{-\gamma}{2}\frac{\partial \mu_2}{\partial x_k} + \frac{\gamma^2}{6}\frac{\partial \mu_3}{\partial x_k}\right), \qquad (4.44)$$

where

$$\delta = 1 + \gamma^2\mu_2/2 - \gamma^3\mu_3/6$$

The quantity δ is proportional to the expected marginal utility of income and is not a constant in general. For small changes in x it can be assumed to be approximately constant—an assumption made in the following analysis.

The terms on the left-hand side of (4.44) would be set equal to zero by the risk-neutral farmer. When the farmer is risk averse, the effects of the input on the higher moments of output also enter the first-order condition. Suppose, for example, that x_k is a risk-reducing input such as a pesticide. Logic suggests that the farmer would utilize more of the input, because of its risk-reducing effect, so that the expected marginal product, $\partial\mu_1/\partial x_k$, would be less than the normalized factor price w_k^* at the optimal solution. The difference between these two terms is given by the right-hand side of (4.44), and is determined by the farmer's degree of risk aversion, represented by γ, and by the derivatives of μ_2 and μ_3 with respect to x_k. The right-hand side equals the risk premium associated with input use, and measures the amount the risk-averse farmer would have to be compensated to use the same input level as a risk-neutral farmer having the same technology.

Let the moment functions be quadratic functions of the variable inputs x and fixed inputs z, so that the moment function derivatives are linear functions of the inputs. Write these derivatives as $\partial\mu_i^j/\partial x_k = \beta_{ik}^j(x^j, z^j)'$ where the superscript $j = a, b$ denotes the two farm groups, β_{ik}^j is a parameter vector comformable to the vector (x^j, z^j), and subscript $k = 1, \ldots, n$ denotes the variable input. Also let $B_i^{j\prime} = (\beta_{i1}^{j\prime}, \ldots, \beta_{in}^{j\prime})$ and $w = (w_1, \ldots, w_n)$. Now (4.44) can be written in matrix form as

$$(B_1^j - \gamma B_2^j/2\delta + \gamma^2 B_3^j/6\delta)(x^j, z^j)' - w^* = 0 \tag{4.45}$$

Partition B_i^j into B_{ix}^j and B_{iz}^j conformably with (x, z). Rewrite (4.45):

$$\begin{aligned}(B_{1x}^j &- \gamma B_{2x}^j/2\delta + \gamma^2 B_{3x}^j/6\delta)x^j \\ &+ (B_{1z}^j - \gamma B_{2z}^j/2\delta + \gamma^2 B_{3z}^j/6\delta)z^j - w^* = 0\end{aligned} \tag{4.46}$$

or as

$$B_x^j x^j + B_z^j z^j - w^* = 0$$

The solution to the expected utility maximization problem is

$$x^{j*} = (B_x^j)^{-1}(w^* - B_z^j z^j), \quad j = a, b \tag{4.47}$$

Several observations are in order about the use of this kind of model for efficiency and welfare analysis. First, since a local approximation to the decision maker's objective function is used to derive the optimal input levels, the model is likely to be valid only for relatively small changes in the model's parameters. For large parameter changes, the model may fail to be a good approximation, and may fail to satisfy the theoretical properties (for example, the objective function may not be concave in inputs). A second, related point is that the estimated model may fail to satisfy the concavity property required for a solution to the maximization problem either locally or globally. Such failures suggest that the model is somehow misspecified, or that there are other empirical problems, like data errors. It is possible to impose curvature properties in estimation, given currently available numerical methods and available software, although it is costly to do so (see, for example, Hazilla and Kopp, 1986). However, if the unrestricted model fails to satisfy curvature and other theoretical properties, it is not clear that a model with curvature imposed is going to provide an accurate representation of the technology.

Once estimates of the expected utility functions have been obtained, it is possible to compare the relative technical efficiencies of the two groups of farms. This can be done at any data point, as at the average factor proportions for each group, for instance. Denoting the average factor proportions of each group as $(\bar{x}^j, \bar{z}^j)$, relative technical efficiency (using the group b technology as numeraire) is given by

$$RTE(\bar{x}^j, \bar{z}^j) = EU^a(\bar{x}^j, \bar{z}^j)/EU^b(\bar{x}^j, \bar{z}^j), \quad j = a, b \tag{4.48}$$

By calculating the *RTE* indexes at various input combinations, it can be determined whether or not one group's technology is uniformly efficient relative to the other's. Note that the analysis of technical efficiency could be made whether or not the theoretical model satisfied curvature conditions with respect to the variable inputs.

If the model satisfies concavity properties, relative allocative efficiency can be calculated for each group by using the estimated technology for that group and the estimates of the optimal variable inputs given by equation (4.47):

$$RAE^j = EU^j(\bar{x}^j)/EU^j(x^{j*}), \quad j = a, b \tag{4.49}$$

The optimal factor demand functions (4.47) also can be used to investigate other aspects of behavior—the effects of risk aversion on input use, for example. The elasticity of demand in the model depends on $(B_x^j)^{-1}$. In the case of one variable input, it can be shown that this term contains the risk-aversion parameter γ and the input's second-order coefficients from the quadratic moment functions. Using this expression, it can be shown that in the risk-neutral case the elasticity of factor demand depends only on the convexity of the first moment function in (4.40), as is true in the neoclassical model; but if the farmer is risk averse, the elasticity generally depends on the degree of risk aversion and on the convexity properties of all of the moment functions in (4.40) and (4.41).

By using the approximate negative exponential model presented in the preceding section, the measures of the rate and bias of technological change (see definitions 7 and 8, chapter 3) can also be computed. The rate of technical change is defined in definition 7 (chapter 3) as one minus the rate of technical efficiency. This computation thus can be made by using (4.48). The bias of technical change was defined in terms of the effect of technical change on factor shares (definition 8, chapter 3). To obtain an expression for this bias, note that with the approximate negative exponential utility function, the stochastic marginal rate of technical substitution (SMRTS) (defined in chapter 3) is:

$$\text{SMRTS}_{i,j} = \left[\frac{\partial \mu_1}{\partial x_i} + \delta^{-1}\left(\frac{\gamma}{2}\frac{\partial \mu_2}{\partial x_i} - \frac{\gamma^2}{6}\frac{\partial \mu_3}{\partial x_i}\right)\right] \Big/ \left[\frac{\partial \mu_1}{\partial x_j} + \delta^{-1}\left(\frac{\gamma}{2}\frac{\partial \mu_2}{\partial x_j} - \frac{\gamma^2}{6}\frac{\partial \mu_3}{\partial x_j}\right)\right]$$

Then, using definition 8 of chapter 3, the estimated SMRTS can be employed to calculate the bias term B_{ij}.

Conclusion

This chapter developed the econometric methods needed to measure stochastic technology, estimate producers' risk attitudes, and analyze the economic efficiency and welfare of farm firms. With data representing the outputs and inputs of a group of producers and the prices they face, these methods make it possible to quantify the technological relationships between production inputs and output as a conditional probability distribution of output given inputs. In estimating this probability distribution, it is necessary to take into account the sequential nature of the manager's decision problem. Once the technology has been estimated, it is possible to measure the distribution of risk attitudes in the producer population represented by the data. Combining the estimates of the technology and the distribution of risk attitudes, it is then possible to evaluate the economic efficiency of producers in the population and to analyze the effects of policies—such as pesticide restrictions or the introduction of integrated pest management technology—on the efficiency and welfare of producers. The following chapter illustrates how these methods can be used by applying them in a case study of California tomato growers.

Notes on the Literature

Chapter 4 is based on the author's recent work. The material on modeling stochastic production in terms of the conditional distribution of output, revenue, or profits is based on Antle (1983b) and Antle and Goodger (1984). The formulation of dynamic econometric production models is based on Antle (1983a, 1984, 1986, 1988), Antle and Havenner (1983), Hatchett (1985), and Antle and Hatchett (1986). The estimation of producers' risk attitudes is based on Antle (1987). The formulation of the model for efficiency and welfare analysis is based on Antle (1985).

The literature treating production processes as conditional distributions began with Day's (1985) analysis of experimental yield data, which used the method of moments (to be distinguished from the moment-based approach discussed here). Subsequently, Anderson (1973) explicitly modeled sample moments of yield distributions as functions of input variables, and discussed the data problems associated with the method of moments. Roumasset (1976) also used this kind of approach. Just and Pope (1978, 1979b) discussed restrictions imposed by econometric production models on the implied relationship between input decisions and production risk in a mean-variance framework, and suggested a heteroscedastic additive-error model to overcome the limitations of the commonly used multiplicative error econometric specifications. Antle (1983b) showed that both the multiplicative error and the Just-Pope additive error models impose restrictions on the higher moments of the distribution, and proposed an econometric method which does not impose such restrictions.

The literature on risk attitude estimation is quite large and diverse. For reviews of the early psychology literature, see Edwards and Tversky (1967) and Coombs, Dawes, and Tversky (1970). More recently, econometricians have proposed random utility models for analysis of discrete choices; see McFadden (1983), for example. In the agricultural economics literature, experimental methods have been used and advocated by Binswanger (1980, 1982) and by Binswanger and Sillers (1983). For reviews of programming and other methods, see Newbery and Stiglitz (1981), Hazell (1982), and Pope (1982).

five

A CASE STUDY: CALIFORNIA PROCESSING-TOMATO PRODUCTION

In chapter 5 the theory and methods developed in the preceding chapters are utilized to measure and analyze the role that pesticides and integrated pest management play in the welfare of California farmers who grow tomatoes for canning and use in processed foods, such as tomato paste. This case study illustrates how the proposed econometric framework can be used to evaluate the effects of pesticide policy on producer welfare, taking into account the effects of sequential decision making and production risk. The results of the case study also are of interest in their own right, as the study provides new evidence on the nexus of production risk, pesticides, and integrated pest management techniques.

The chapter begins with background material on processing-tomato production and on an IPM program that was introduced in 1984, on which the case study is focused. Subsequent sections of the chapter discuss the case study data and the specification of the econometric model, present and interpret the econometric results from estimation of the technology and risk attitudes, and use those results to analyze the effects of pesticide policy and the IPM program on producer welfare.

Background

Processing tomatoes are a major crop in California's Central Valley. California produces about 90 percent of the annual U.S. crop, and about 65 percent of the state's crop is grown in the Sacramento Valley, which is the northern part of the Central Valley and is the focal area of this study. In 1984 there were nearly 300,000 acres planted to processing tomatoes in the United States, of which 244,000 were in California. The yield in California averages more than 25 tons per acre, in comparison to the 14 ton-per-acre average in the rest of the country. California's 1984 production was more than 6.5 million tons, which was valued at about $325 million, or about $50 per ton.

Processing tomatoes pass through growth stages during a season, typically categorized as seedling establishment, vegetative, flowering, and fruiting. Each stage involves differing nutrient requirements and different kinds of pest problems. While the earlier stages are important to the total tomato fruit yield, the fruiting stage largely determines tomato quality. In the fruiting stage, the tomato fruitworm and beet armyworm are perennial problems in the Sacramento Valley. The larvae of both species can damage fruit by entering through the skin, but the tomato fruitworm is particularly damaging because its excrement contaminates the fruit. These pests are particularly troublesome later in the season. As a result, many growers have attempted to avoid pest damage by planting earlier crops. However, because of the constraints of weather and labor and machinery availability, most growers in the Sacramento Valley cannot avoid planting later in the season when these pests can cause substantial damage.

Fruit damage affects growers economically in several ways. Each load of tomatoes (of about 50 tons each) is sampled at state inspection stations for compliance with a California state quality standard that stipulates a 2 percent damage limit on each load accepted by processors. Loads which exceed the standard must either be re-sorted by the grower—a costly process—or otherwise disposed of. The quality standard provides growers with a strong incentive to ensure that very few, if any, loads are rejected. This policy can thus be interpreted as creating an incentive for growers to use insecticides when they are judged necessary to ensure that a crop will meet the state standard.

It would be wrong to suggest, however, that without the state standard growers would not have an economic incentive to minimize fruit damage. Processors encourage growers to deliver loads with less than 2 percent damage by providing price premiums for high-quality loads and by awarding contracts to growers who have a record of providing good-quality fruit in previous seasons. As a result of these incentives, loads delivered by growers to processors typically have very low damage levels (often of only trace amounts of less than 0.5 percent).

Growers are also concerned about fruit quality because much of the damaged fruit drops off the plant before harvest or is rejected in the harvesting process, resulting in lower yields rather than higher postharvest damage. An unpublished study by the University of California Statewide IPM Program found that most of the fruit that was damaged was lost before or during harvest, and did not show up as reduced postharvest quality.

The IPM Program for Tomatoes and the Case Study

In 1984 the University of California Statewide IPM project introduced a program for reduction of fruit damage due to lepidopterous caterpillars (primarily fruitworm and beet armyworm) on about 2,000 acres of processing tomatoes in the Sacramento Valley. This IPM program was based on research by University of California entomologists, who devised a probability-based method of sampling fields for fruit damage and the presence of caterpillar eggs, and decision rules for insecticide spraying based on the sample results. To use this program successfully, each field must be sampled weekly. When the program is being followed, growers are making insecticide spray decisions sequentially. Typically, growers who do not take part in this IPM program use informal methods for making decisions on spraying, based on visual field inspections, on the advice of pest control consultants, and on their experience. In testing the entomological and economic viability of the program in the Sacramento Valley, cooperating growers and their pest management consultants volunteered to be trained in the use of the sampling technique and to apply it on some of their fields. The program is described in detail in Park (1985) and in Weakley, Zalom, and Wilson (1986); other relevant information is available in a manual on IPM for tomatoes published by the University of California Statewide IPM Program.

The economic benefits and costs of the IPM program are easily understood. They can be categorized as: changes in expected yield and revenues resulting from the program on mean fruit damage; changes in pest management costs, primarily related to the number of sprays, the kind of spray materials used, and labor for field monitoring; and changes in production risk resulting from effects of the program on the risk of fruit damage.

The unpredictability of fruit damage from pests means that fruit damage represents an important aspect of production risk for tomato growers. The tomato IPM program is designed to offer growers a way to increase mean yields and revenues, as well as to reduce production risk, without using more chemicals. If growers spray less when they take part in the program, they should be able to reduce insecticide costs. On the other hand, there are additional costs, direct and indirect, associated with the program. The direct costs to growers result from a possible increase in the labor involved in monitoring the fields systematically. An indirect cost might be incurred from an increase in production risk if the program is not used correctly, or if the design of the program is faulty. Beyond these private benefits, the IPM program could have broader beneficial effects on pest resistance, biological control, and the environment if it leads to a reduction in overall pesticide use.

Description of the Case Study Data

The production data were collected by Park and the author in interviews with growers who agreed to provide information about their production in 1983 and 1984; a listing and detailed description of the data are available in Park (1985). The data were collected on a field-specific basis, the average field size being 75 acres. Summary statistics for the data and the sample of growers are given in appendix A to this chapter. The final sample for both output and input data consisted of 21 fields that were in the IPM program in 1984, and 85 other fields that were in the program in 1983 and 1984. The damage data were collected for 56 fields in 1984. Each field was sampled shortly before it was harvested. In addition to these quantity data, tomato output prices, labor wage rates, and pesticide prices were obtained from growers and pesticide dealers. The price data were used to identify the structural model and to quality-adjust the insecticide data, as described below.

As the group of participants selected for the study was not a random sample of the producer population, a potential problem of a sample selection bias exists. However, since the unit of observation was the individual field and not the farm, this source of bias is judged not to be important. Fields included in the study, either as fields in the IPM program or as "control" fields not in the IPM program, were randomly selected from each farm in the study. It should be emphasized that the decision to use or not use the IPM technology on a given field was made by the researchers, not by the farmer. Thus, in comparing statistically the IPM and non-IPM technologies estimated from the two groups of fields, the problem of potential sample selection bias resulting from the farmers' choice of technology does not arise.

Ten quantity variables were utilized in the econometric analysis. The variables included acreage, fertilizers, irrigation water, insecticides, worm damage, net final tomato output, and dummy variables. All of these were obtained from farmers' field-specific, written production records. Precise output and insecticide data were obtained from California state grading sheets and from the detailed pesticide records that growers are required to keep. Of the three dummy variables used in the analysis, the IPM variable indicated that a field was in the IPM program. Two other dummy variables indicated when a field was planted.

Several important variables were omitted from the analysis, primarily because growers did not keep detailed written records for these data. Machinery utilization data could not be obtained (either from written records or from memory recall) largely because such a variety of machinery was used on the farms for land-preparation and cultivation operations. Only a few growers kept detailed labor records by field. Most field-level labor input figures were obtained from growers' recall, and were of questionable accuracy. Since the field-preparation and cultivation technology is remarkably uniform in the study area, the mechanical and labor inputs were used in the analysis in approximately fixed proportion to acreage, so they could be omitted from the model without introducing large biases. More important to this study was the lack of accurate data on labor devoted to pest monitoring activities. Such information was collected, but was not utilized in the econometric analysis because of the apparent inaccuracy of grower recall.

Variation in the quality of insecticides is a major problem in the analysis of pest management technology. The use of both highly concentrated insecticides, which require very low application rates, as well as the use of materials requiring much higher application rates for the same degree of pest control, means that there are large quality differences in the materials used. These differences are reflected, to varying degrees, in insecticide prices. One solution to the quality-difference problem is to use experimental measurements of each material's efficacy under local conditions to quality-adjust the quantity data. However, as such information was not available for this study, a hedonic regression technique was utilized instead (see Freeman, 1979, chapter 4 for a discussion of the hedonic method). Underlying the hedonic method is the concept that market prices reflect quality differences associated with qualities of the good. By regressing price on indicators of quality, the price component (or implicit price) associated with each quality component is measured. This implicit price can be used to quality-adjust the quantity data into a homogeneous variable measuring pesticide input in standard efficiency units.

In insecticide treatments for fruitworms and beet armyworms in processing tomatoes, one measure of the quality of the insecticide is the quantity of active ingredient applied per standard treatment per acre. Since such information was not available in the data collected for this study, quality was measured in terms of the planned or expected quantity applied per acre, where the planned quantity was obtained from the regression of actual quantity applied on the predetermined variables that represented the a priori information used by the farmer in planning production. Hedonic price regressions were run for insecticide prices, using planned applications and the IPM and planting-date dummy variables as explanatory variables. The results of the model, with both linear and quadratic terms for pesticide applications, were as follows; t-statistics are in parentheses:

$$\begin{aligned} ISP = {} & \underset{(5.287)}{26.204} - \underset{(-2.450)}{13.219} \cdot IF + \underset{(1.822)}{1.702} \cdot IF^2 \\ & + \underset{(1.393)}{4.998} \cdot IPM + \underset{(3.037)}{7.111} \cdot D2 + \underset{(1.005)}{3.165} \cdot D3 \end{aligned}$$

$$R^2 = 0.213 \qquad F = 5.411$$

where ISP = insecticide price, IF = planned insecticide application, IPM = dummy variable (1 if a field is in the program, 0 otherwise), $D2$ = midseason planting dummy (1 if midseason, 0 otherwise), and $D3$ = late season planting dummy (1 if late season, 0 otherwise).

The results for the cubic model were:

$$\begin{aligned} ISP = {} & \underset{(4.617)}{26.441} - \underset{(-1.503)}{13.831} \cdot IF + \underset{(0.549)}{1.982} \cdot IF^2 - \underset{(-.077)}{.033} \cdot IF^3 \\ & + \underset{(1.346)}{4.935} \cdot IPM + \underset{(2.859)}{7.162} \cdot D2 + \underset{(0.883)}{3.280} \cdot D3 \end{aligned}$$

$$R^2 = 0.231 \qquad F = 4.464$$

The two sets of results are similar, and both indicate a statistically significant negative relationship between the planned insecticide application rate IF and insecticide price ISP. The positive coefficient on IF^2 indicates that the relationship between ISP and IF is negative, but at a declining rate. These results show that a higher-quality material (having a lower-planned application rate) has a higher implicit price. The coefficients on the dummy variables indicate that higher-quality materials were applied on fields in the IPM program and on late-planted fields where growers expected more worm damage.

The Damage Model

In processing-tomato production, IPM and other pest management practices designed to control damage from caterpillar pests are directed at the fruiting stage of plant growth. Given the well-known statistical problems that arise in econometric analysis with nonexperimental data (see Leamer, 1978, for example), it is reasonable to suggest that the most accurate econometric measurement of pest management technology should be obtained by measuring the direct outcomes of pest management activities. Consequently, it seems desirable to focus econometric analysis on preharvest fruit damage rather than on final output. However, to do so requires that several assumptions be made. The present section discusses those assumptions, and outlines the resulting modifications of the general econometric methods that were presented in the preceding chapter.

In order to investigate fruit damage separately from gross output (functional separability has been discussed in chapter 4), it is necessary to invoke a separability assumption. For this purpose, let the relation between realized postharvest output q, gross or preharvest output q^g, and the percentage damage D be

$$q = H(q^g, D),$$

where H represents the "harvesting production function" that depends on preharvest output and the damage rate (other harvesting inputs are omitted for simplicity). Assuming that much of the damaged fruit is rejected in the harvesting process, the function H can be specialized to

$$q = q^g(1 - D) \tag{5.1}$$

This relationship shows that postharvest output q is a random variable resulting from the unpredictable influence of weather, pests, and other factors beyond the farmer's control that affect q^g and D. Thus the distribution of q is determined by the joint distribution of q^g and D. The latter distribution can be written $f[q^g, D|x_m, x_g]$ where x_m denotes pest management inputs and x_g is a vector of other inputs.

The fact that fruit damage from caterpillar pests occurs toward the end of plant growth suggests that most random events (weather, for example) influencing gross yield occur before the fruiting stage. This in turn suggests that it may not be unreasonable to assume that q^g and D are statistically independent phenomena. Furthermore, that inputs into other production operations are distinct from pest management inputs suggests that x_g affects the distribution of q^g, but not D. Therefore, the joint distribution of q^g and D can be factored into two parts so that

$$f[q^g, D|x_m, x_g] = f_g[q^g|x_g]\ f_m[D|x_m] \tag{5.2}$$

This is the condition for weak stochastic separability defined in chapter 4. The assumption of stochastic separability is important because it means that it is possible to analyze pest management directed at fruit damage in terms of the damage distribution.

In this case study, Sacramento Valley tomato production is analyzed separately from other production activities. Some producers, however, grow several crops in the same summer season, which raises the problem of joint production. On the basis of the technological requirements of these crops, it can be argued that tomato production is technologically nonjoint with respect to other crops grown. Recall that for exact stochastic nonjointness to hold, distinct inputs must be allocated to each process—inputs which do not affect the productivity of other processes—and the correlations across crops must not be functions of each crop's inputs. While some capital may be used jointly in production with other crops, by and large it is reasonable to assume that distinct fixed and variable inputs are allocated to tomato production. And while tomatoes and other crops are subject to the same weather, crops are planted at different times and thus are affected differently by random weather events. Moreover, except in the early spring when some crops are planted and late fall when some are harvested, the arid climate is quite stable and the tomato crop is irrigated. These conditions suggest that to the extent that crop yields are correlated, the correlations are not a function of field-specific inputs, but rather of the particular relation of each crop to the seasonal weather patterns. Thus the nonjointness assumption may be a reasonable one. It is assumed in the following analysis that the tomato production technology satisfies the conditions for exact stochastic nonjointness given in chapter 4.

The unit of observation in this case study is the field. A field is a distinct parcel of land that is planted to tomatoes and managed as an economic unit. Generally, each field is planted at a specific time, and most management decisions about cultivation, irrigation, fertilization, pest management, and harvesting are field-specific. In interpreting the field as an economic unit, two concepts of scale are relevant. First, economies or diseconomies of scale could exist as a function of field size. Pest management decisions, for example, might be applied more efficiently to larger fields than to smaller ones. Second, in aggregating from one field to a whole-farm operation, there may also be scale economies. That competing farms in the Sacramento Valley range in size from a few hundred acres to thousands of acres suggests, however, that there are effectively constant returns to scale at the farm level. Thus while there may be scale effects related to field size, economies of scale at the farm level do not appear to be important.

A fundamental question underlying the analysis of firm behavior and welfare is the form of the producer's objective function. Very little empirical information is available to guide the researcher in the specification of the objective function. For the purposes of this study, which is cast within the expected utility maximization framework, it will be assumed that the processing-tomato grower's objective function within the growing season is the maximization of expected utility of net returns to fixed factors of production (producer surplus). In this formulation, fruit damage affects utility only through its influence on net returns. While it is possible that damage in one season may affect a grower's ability to win a contract in a future season, such a dynamic effect is not considered in the present analysis.

Assuming that preharvest damage is translated into postharvest yield reduction, net returns are defined as

$$\pi = pq^g(1 - D) - wx$$

Furthermore, since gross output is largely determined in early growth stages while most fruit damage is determined just before harvest, q^g can be viewed as a random variable that is realized and observed by growers when pest management decisions for fruit damage are made. As product price p is predetermined in contracts with processors, the term pq^g can be viewed as predetermined relative to pest management decisions including fruit damage. Therefore, net returns can be normalized by pq^g:

$$\pi^* = (1 - D) - w^*x \quad \text{where } \pi^* = \pi/pq^g \text{ and } w^* = w/pq^g \tag{5.3}$$

It should be noted that in this form net returns are expressed as a percentage of pq^g. The quantity $(1 - D)$ can be interpreted as tomato quality, and the damage rate D can be interpreted as a "negative output."

Using (5.3), the moments of profit can be seen to be closely related to the moments of damage. Letting the moments of damage be μ_i, $i = 1,2,3$, mean profit is $1 - \mu_1 - w^*x$, the variance of profit equals μ_2, and the third moment of profit equals $-\mu_3$. Thus the expected utility of profit can be expressed as

$$EU = u[1 - \mu_1 - w^*x, \mu_2, \mu_3] \tag{5.4}$$

Assuming that the decision maker is both Arrow-Pratt and downside risk averse, the expected utility function should satisfy

$$\frac{\partial u}{\partial \mu_1} < 0, \quad \frac{\partial u}{\partial \mu_2} < 0, \quad \frac{\partial u}{\partial \mu_3} < 0$$

Within this decision-making framework, the flexible moment-based approach developed in the previous chapter can be utilized to model the stochastic technology as a function of the moments of the damage distribution. Note that since damage can be thought of as a negative output, the interpretation of the odd moments (first, third, and so on) is the opposite of the effect that the moments of usual output have on expected utility.

In order to implement the welfare and efficiency analyses discussed in chapters 3 and 4, it is necessary to specify a utility function. Generally, little a priori information is available about the parametric form of decision makers' utility functions. For this study the approximate negative exponential utility function, introduced in chapter 4, was employed. Given (5.4), it is specified as

$$\begin{aligned} EU(\pi^*) = 1 &- \exp(-AP(1 - \mu_1 - w^*x)) \\ &- \exp(-AP(1 - \mu_1 - w^*x))(AP^2\mu_2/2 \\ &+ AP\cdot DS\mu_3/6), \end{aligned} \tag{5.5}$$

where the μ_i are the central moments of the distribution of D, and $AP = -U''/U'$ is the Arrow-Pratt absolute risk aversion coefficient, and $DS = U'''/U'$ is the absolute downside risk aversion measure introduced in chapter 4. Special note should be made of the term involving the third moment and its coefficient $AP\cdot DS$. If AP risk aversion is constant, then $AP^2 = DS$ and the function is a third-order Taylor series approximation to the negative exponential utility function. This specification allows for the fact that AP risk aversion may not be constant, in which case the effect of the third moment on expected utility may be greater or smaller than in the negative exponential case. This utility function can be implemented by using the econometric methods discussed in chapter 5 to estimate the terms AP and DS (see equations 4.30–4.32).

One alternative to the approximate negative exponential utility function which does not require the Taylor series assumption is the use of econometric risk attitude estimates to approximate the derivatives of the expected utility function. Thus an alternative approximate expected utility function which is strongly separable in the moments would be:

$$EU = g(\mu_1 + r_2\mu_2 + r_3\mu_3), \quad g' > 0, \tag{5.6}$$

where the r_i, $i = 2,3$, are proportional to the derivatives of the expected utility function as defined in equation (4.32). Here expected utility is defined in the units of mean normalized net returns, because that is the form in which the derivatives of the expected utility function are estimated. To avoid one of the restrictive assumptions of the expected utility theory (the "independence axiom"), this model can be modified along the lines of the model proposed by Machina (1982), as discussed in appendix B to this chapter. Although the empirical results reported below are based on the approximate negative exponential utility function, it should be noted that (5.6) also was used with the Sacramento Valley data, with similar results. These results suggest that neither of these specifications biases the results, at least in relation to the other model.

For the estimation of the stochastic technology, the moments of the damage distribution were specified as quadratic functions of the following form:

$$\begin{aligned}\mu_i = \beta_0^i + \beta_1^i W + \beta_2^i I + .5\beta_3^i W^2 \\ + \beta_4^i W \cdot I + .5\beta_5^i I^2 + \beta_6^i I \cdot IPM + \beta_7^i A + \beta_8^i IPM \qquad (5.7) \\ + \beta_9^i D2 + \beta_{10}^i D3, \quad i = 1, 2, 3\end{aligned}$$

where W is water applied, I is quality-adjusted insecticide, A is acreage, IPM is a dummy variable (1 if a field was in the program, 0 otherwise), $D2$ is the midseason planting dummy variable (1 if a field was planted in midseason, 0 otherwise), and $D3$ is the late-season planting dummy variable (1 if a field was planted in late season, 0 otherwise).

The assumption of weak stochastic separability between gross output and the damage rate provides the justification for including only water and insecticides as variable inputs in the quadratic specification. Other inputs, such as fertilizer and cultivation, are assumed not to affect the damage rate. Acreage is included linearly to account for a possible effect of field size on damage. The interaction term between the insecticide variable and the IPM program dummy variable is introduced to measure the effects of participation in the program on the productivity of pesticides. The planting-date dummy variables are introduced to measure the effects of early, middle, and late planting on damage.

Econometric Results

The Damage Distribution Moments

The parameter estimates of the quadratic damage distribution moment functions, as specified in equation (5.7), are presented in table 5–1. These estimates were obtained by using the nonlinear two-stage

Table 5–1. Nonlinear Two-Stage Least Squares Estimates of the Damage Distribution Moments for Sacramento Valley Processing-Tomato Producers

	Moment					
	First		Second		Third	
	coef.	*t*-stat.	coef.	*t*-stat.	coef.	*t*-stat.
Intercept	2.376	2.677	6.122	1.774	17.523	1.359
Water	.113	.094	4.022	1.119	19.079	1.444
Insecticide	.586	.362	−7.272	−1.451	−31.341	−1.872
Water^2	.166	.864	−.234	−.335	−2.054	−.790
Water × Insecticide	.084	.219	.376	.255	−.240	−.044
Insecticide^2	−.146	−.308	3.109	1.823	13.156	2.057
Acres	−1.970	−1.575	−6.208	−2.350	−18.942	−1.900
IPM	−1.165	−.991	2.775	.735	15.787	1.038
IPM × Insecticide	−.004	−.005	−4.164	−1.717	−16.987	−1.680
Midseason planting	1.010	1.200	2.619	.966	6.608	.633
Late planting	.356	.398	−2.399	−.786	−13.776	−1.198
X^2	62.595		80.179		33.139	

Sample size = 106

least squares procedure, in order to account for sequential decision making in pesticide application. All three functions are statistically significant, as judged by the X^2 statistics at the bottom of the table. However, many of the individual coefficients have relatively large standard errors, and are thus not precisely estimated. Because of this lack of precision, the results must be interpreted carefully.

In order to interpret the parameter estimates with regard to the effects of pesticides on damage distribution, it is useful to compute the marginal effects of the input on each moment. The derivatives of the moment functions are

$$\partial\mu_i/\partial I = \beta_2^i + \beta_4^i W + \beta_5^i I + \beta_6^i IPM \tag{5.8}$$

Thus the elasticity of the moments with respect to inputs (defined in equation 4.35) can be expressed as functions of the parameters of the model. These elasticities are useful in interpreting the results, and are presented in table 5–2.

Observe that the derivative in equation (5.5) is a function of the parameter β_6^i which measures the interaction between insecticides and IPM. Since the IPM program is designed to improve the farmer's

Table 5–2. Mean Elasticities of Damage Distribution Moments

	Moment		
	First	Second	Third
Acreage	−1.252	−3.679	−7.673
Water	.344	2.401	5.273
Insecticide	.248	−1.722	−4.701
Insecticide (non-IPM)	.294	−.411	−1.878
Insecticide (IPM)	.247	−3.264	−8.020
IPM	−.261	−.291	−.172

Note: Elasticity computations are based on parameter estimates in table 5–1 and equation (4.35). Non-IPM indicates the computation was based on the estimated technology with the IPM dummy variable equal to 0. IPM indicates the computation was based on the estimated technology with the IPM dummy variable set equal to 1.

ability to use pesticides in a timely and effective manner, it is hypothesized that this parameter is negative, thereby indicating that those fields in the program should have lower mean damage and lower damage risk than those fields not in the program. In table 5–1, one can see that this hypothesis cannot be rejected for the second and third moment functions at the 10 percent significance level, although the coefficient in the mean equation is small and has a large standard error. These results suggest that the IPM technology does enhance the effectiveness of pesticide use, and that the increased effectiveness comes about primarily through the risk attributes of the technology, and not through an increase in the mean marginal product of insecticides. However, these findings do not necessarily imply that the IPM program leads farmers to use fewer pesticides, because there are both substitution and output effects of this productivity change. It is necessary to evaluate the effects of the program on pesticide demand to know which of these effects dominates.

The parameters on the planting dummy variables also have implications for technical efficiency. Under the null hypothesis that later planting involves heavier pest infestations as compared to early planting, the coefficients of the dummy variables in the mean equation should be positive, as they are. If planting in the later season also involves higher damage variability, the dummy coefficients should be positive for the second and third moments. This holds true for the midseason dummy, but not for the late-planting dummy (although these coefficients are not accurately estimated and must be interpreted cautiously).

The coefficients of the acreage, water, insecticides, and IPM variables can be interpreted in terms of the elasticities given in table 5–2, which shows the elasticities of each variable at the sample mean of the data. The insecticide elasticities are also presented at the means of the data, but with the IPM dummy set at the values of zero and one to represent the shift in the technology resulting from use of the IPM program. The acreage elasticities indicate that there may be economies in pest management according to field size, as the larger fields appear to be associated with lower mean damage and less damage risk. The elasticities also show that increased application of irrigation water leads to higher mean damage and higher damage risk, a finding which could be interpreted in terms of the interaction of irrigation and the pest populations. Insecticides are hypothesized to be risk-reducing, and this is evidenced in the results by the negative elasticities with respect to the second and third moments. Moreover, the risk-reducing effects are much higher for the fields in the program than for those that were not, indicating that the IPM program was successful in increasing the effectiveness of pesticides in reducing risk, presumably by improving the timing of pesticide applications in relation to the pest populations. The IPM elasticities show that the fields in the IPM program have less mean damage and less damage risk, as all three moments of the damage distribution are decreasing in the IPM variable. Thus the technology estimates indicate that the IPM program increases the effectiveness of insecticides and has an overall beneficial effect on the damage distribution.

Note also that table 5–2 shows insecticides to have a positive effect on mean damage, a finding that could be considered surprising and counterintuitive. This finding is not necessarily inconsistent with the behavior that would be exhibited by a rational, risk-averse decision maker (although it would be inconsistent with the behavior of a risk-neutral decision maker), because a rational risk-averse decision maker may utilize a risk-reducing input in such a way that the mean marginal product is negative. This would occur when the marginal risk premium was sufficiently large and positive so that, in order to satisfy the first-order condition for expected utility maximization, the expected value of the marginal product would have to be less than the factor price, and could be negative (see the discussion following equation [3.8]). Since insecticide is found to be risk-reducing, a negative mean marginal product of insecticide cannot be ruled out a priori. We cannot conclude whether insecticide is being efficiently used without measuring the grower's objective function and evaluating the efficiency of input use, as is done in the following sections.

To investigate the need to account for sequential decision making, the moment-based model was also estimated by using a procedure which did not take into account the endogeneity of insecticides. The resulting estimates yielded implausible results: the elasticities of the insecticide variables with respect to the second and third moments showed that insecticides were risk-increasing, rather than risk-reducing. These findings bear out the importance of accounting for the effects of sequential decision making on the endogeneity of pest management inputs.

Risk Attitude Estimation

The distribution of risk attitudes was estimated by use of the econometric methods described in chapter 4. As explained there, a variety of assumptions can be made concerning the statistical distribution of risk attitudes in the producer population. For the purposes of this investigation, it was assumed that risk attitudes are independent across observations. While arguments can be made both for and against this assumption, it is a practical fact that this assumption is the only one that is feasible without longitudinal data. In interpreting the results from this study's sample, it must be remembered that the sample was not a random one designed to be representative of the entire population of Sacramento Valley tomato growers. Thus it remains an open question whether the results obtained here for risk attitudes are indeed generalizable to the population.

The two-stage least squares version of the instrumental variables technique was utilized to purge the estimates of bias resulting from the dependence of input choices on risk attitudes. To obtain the desirable properties of the estimator, the variables used for instruments must be correlated with insecticide input decisions, but uncorrelated (in the limit) with risk attitudes. The instruments were output and input prices, field acreage, participation in the IPM program, and planting dates. The parameter estimates of the risk attitude distribution parameters are presented in table 5–3. This table gives estimates of the full three-moment model (Model 1 in the table), as well as of two submodels which included the first and second and first and third moments only (Models 2 and 3). The latter two models were used because it was found that the three-moment model produced implausible results, apparently as a result of the high degree of correlation between the second and third moment derivatives. The two-moment models, in contrast, produced plausible results. The justification for the use of Models 2 and 3 is that the estimates from Model 1 are highly unreliable because of the extreme degree of multicolinearity. However, it is also well known that the estimates of Models 2 and 3 are likely to be biased upward (in absolute terms) as a result of the exclusion of a variable which is highly positively correlated

Table 5–3. Two-Stage Least Squares Estimates of Risk Attitude Distribution Parameters

	Model		
	1	2	3
Mean Risk Attitudes			
$\theta_{\bar{1}}^{2}$	1.784	−.345	
	(3.847)	(−5.846)	
$\theta_{\bar{1}}^{3}$	.474		.085
	(4.429)		(6.415)
Covariance Parameters			
θ_0	3.005	20.093	4.402
	(2.045)	(4.728)	(2.433)
θ_{22}	.864	−.067	
	(.956)	(−.646)	
θ_{33}	.036		.000
	(.864)		(.169)
$2\theta_{23}$	.348		
	(.889)		

Note: *t*-statistics are given in parentheses.

with the included variable. Thus the results of Models 2 and 3 in table 5–3 should be interpreted carefully.

The first notable observation about the results for Models 2 and 3 in table 5–3 is that the estimate of $\theta_{\bar{1}}^{2}$ is negative and the estimate of $\theta_{\bar{1}}^{3}$ is positive, indicating that expected utility is decreasing in the variance and increasing in the third moment of net returns. Using the Taylor series approximation to the utility function, this finding implies that the population is characterized by both Arrow-Pratt and downside risk aversion. The second notable result in table 5–3 is that the estimates of the variance parameters θ_{22} and θ_{33} are small and statistically insignificant. This finding indicates that there is no evidence of highly heterogeneous risk attitudes; that is, it suggests that it may be reasonable to use the estimates of risk attitudes of the population mean to represent the population. Using the estimates of the variances and covariance parameters, the correlation between Arrow-Pratt and downside risk aversion is positive. The estimate of the correlation coefficient from Model 2 is .53, and the estimate from Model 3 is .71. While these correlations are based on parameter estimates that have large sampling errors and should be interpreted cautiously, they suggest that growers who are more Arrow-Pratt risk-averse also tend to be more downside risk-averse.

In order to interpret the implications of the parameters in table 5–3, various measures of risk attitude characteristics are presented in table 5–4. The average relative risk premium of about 11 percent suggests that the growers represented in the study may be interpreted as being "moderately" risk averse, in the sense that they would be willing to pay at most about 11 percent of expected returns to insure against risk. The marginal risk premium indicates the effect that a change in insecticide use has on the risk premium. These values are negative, indicating that the input is marginally risk-reducing. The marginal risk premiums calculated for the growers using the IPM technology (the IPM dummy set equal to one in the model) are about double the value at the sample mean, whereas the marginal risk premiums for the non-IPM technology are near zero, indicating that the IPM program greatly increased the effectiveness of insecticides in reducing production risk. Finally, the computations suggest that the population, at the mean, is characterized by both weak and strong decreasing absolute risk aversion. It should be noted, however, that the calculations show only a small deviation from constant absolute risk aversion. Considering that the parameters are estimated with error, the evidence suggests that it would not be possible to reject the hypothesis of constant absolute risk aversion.

In concluding this section, several findings of the econometric estimations deserve emphasis. First, the IPM technology was found to enhance the effectiveness of pesticide utilization, primarily through its effects on the risk attributes of the technology. Consequently, the marginal risk premium of insecticides was near zero for the non-IPM technology, but negative for the IPM technology.

Table 5–4. Risk Attitude Characteristics of Sacramento Valley Tomato Growers

Risk aversion measures	
mean absolute Arrow-Pratt	.690
mean partial Arrow-Pratt	.345
mean absolute downside	.510
mean partial downside	.128
Risk premiums	
average relative	.107
marginal relative	
mean	−.118
IPM	−.234
non-IPM	−.020
Decreasing absolute risk aversion (at population mean)	
weak	Yes
strong	Yes

Note: Material is based on Models 2 and 3 of table 5–3.

Second, accounting for sequential decision making was critical in obtaining reliable estimates of the technology, as was the resulting endogeneity of insecticide input.

Third, the producer population was found to be both Arrow-Pratt and downside risk averse, and no evidence of heterogeneous risk attitudes was found. This result probably reflects the homogeneity of the producer population, and the fact that the farmers in the sample were not randomly selected. It remains an open question whether this sample was representative of the population of tomato growers in the Sacramento Valley.

Analysis of Efficiency and Welfare

By using the econometric estimates of the stochastic technology and of producers' risk attitudes, as well as the approximate utility functions of equation (5.5), the analysis of efficiency and welfare can proceed along the lines outlined in chapter 4. In view of the possible biases in the risk attitude estimates, and in order to explore the effects of differing degrees of risk aversion on the analysis, four models, corresponding to differing degrees of risk aversion, were used to analyze efficiency:

(A) $AP = .2$, $DS = .15$, implying a relative risk premium of 2.8 percent
(B) $AP = .4$, $DS = .3$, implying a relative risk premium of 5.6 percent
(C) $AP = .8$, $DS = .6$, implying a relative risk premium of 11.7 percent
(D) $AP = 1.2$, $DS = .9$, implying a relative risk premium of 16.7 percent

It can be seen that in relation to the results for the population mean presented in table 5–3, models (A) and (B) represent a relatively low degree of risk aversion, model (C) is close to the estimated population mean, and model (D) is about 50 percent higher than the population mean.

Certainty equivalents measured in dollars for a 75-acre field, based on each of the four models and computed for the IPM and non-IPM technologies at the average factor proportions of the IPM and non-IPM groups, are given in table 5–5. These equivalents were obtained by calculating the value of expected utility according to (4.5), then translating into dollar terms, using the inverse function of the negative

Table 5–5. Dollar Certainty Equivalents for IPM and non-IPM Producers

Model	At IPM factor proportions		At non-IPM factor proportions	
	IPM	non-IPM	IPM	non-IPM
(A)	48,140	46,372	48,745	47,489
(B)	47,319	44,962	48,178	46,835
(C)	45,706	42,223	47,058	45,546
(D)	44,154	39,659	45,967	44,296

Note: Based on the assumption of a negative exponential utility function and the utilities from the approximate negative exponential utility function. Models (A) to (D) represent increasing degrees of risk aversion, as defined in the text.

exponential utility function, and assuming constant *AP* risk aversion. Thus for the utility function $U(\pi) = 1 - \exp(-\gamma\pi)$, the certainty equivalent is $CE = -ln[1 - EU(\pi)]/\gamma$. These *CE*s provide the basis for computing the efficiency indexes developed in chapter 3 and for evaluating the welfare effects of IPM adoption and of pesticide restrictions.

The following analyses differentiate the efficiency and welfare of fields in and not in the IPM program. The planting-date variables for the IPM and non-IPM groups (see appendix A) show that the IPM fields generally were planted later in the season, when pest problems were more severe. Thus we can interpret the differences between the IPM and non-IPM factor proportions as reflecting the differences associated with early- and late-season crops.

Alternative interpretations of the welfare and technical efficiency effects of the IPM program are provided in tables 5–6 and 5–7. The technical efficiency indexes in table 5–6, which are all less than one (the IPM technology is the numeraire in the index calculations), show that the IPM program had a positive effect on welfare. In the table, the TE(mean) index accounts only for the effect of the IPM program on mean damage, whereas TE takes both the mean and risk effects of the program into account. The IPM program is seen to have had a greater effect on welfare at the IPM factor proportions. Because of the risk-reducing effects of the program, as the degree of risk aversion increases, the expected utility associated with the non-IPM technology declines relative to the IPM technology. At the non-IPM factor proportions, the contribution of the mean damage reduction is approximately the same as at the IPM factor proportions, but the contribution of risk is much less important. Consequently, the value of the program does not increase as rapidly with the degree of risk aversion. This

Table 5–6. Technical Efficiency Indexes of the IPM Program

	Factor proportions			
	IPM		non-IPM	
Model	TE	TE(mean)	TE	TE(mean)
(A)	.965	.988	.975	.998
(B)	.954	.977	.975	.997
(C)	.936	.958	.973	.997
(D)	.921	.942	.972	.992

Note: TE is technical efficiency as defined in equation (4.48). TE(mean) is the technical efficiency difference resulting only from the effect of the IPM technology on the mean of the damage distribution. Models (A) to (D) represent increasing degrees of risk aversion, as defined in the text.

Table 5–7. Value of IPM Adoption

	Contribution of:		
Model	Mean damage reduction	Damage risk reduction	Total
(A)	1,170	598	1,768
	16	8	24
(B)	1,170	1,187	2,357
	16	16	31
(C)	1,170	2,312	3,482
	16	31	46
(D)	1,170	3,325	4,495
	16	44	60

Note: Based on certainty equivalents in table 5–6 for the IPM group. The first number is the total dollar amount, the second number is the dollar amount per acre. Models (A) to (D) represent increasing degrees of risk aversion, as defined in the text.

difference can be explained by the fact that the non-IPM fields generally were planted earlier in the season when pests were less of a problem, and when relatively fewer insecticides were used (as measured in efficiency units).

Table 5–7 can be used to evaluate the welfare effects of the IPM program in dollar terms. The data show that at low degrees of risk aversion, the value of the reduction in mean damage was about $16 per acre. The importance in taking risk into account is evident at the IPM factor proportions. Even at a relatively low degree of risk aversion, the contribution of the risk-reducing effects of the program is valued at $8 per acre; this value rises rapidly with the degree of risk aversion to $44 per acre. Thus both tables 5–6 and 5–7 show that *the welfare effects of the IPM program would be seriously underestimated if the risk-reducing effects of the program were ignored.*

Table 5–8 presents the allocative efficiency (AE) indexes, the optimal insecticide input levels implied by the model, the certainty equivalents evaluated at the optimal insecticide level, and the dollar cost of allocative inefficiency implied by the certainty equivalents. At the IPM factor proportions, the optimal input level is greater than the actual level by about 25 percent. As risk aversion increases, the optimal input level increases, as would be expected when insecticide is a risk-reducing input, and consequently the AE index declines. The changes in certainty equivalents indicate that the cost of inefficiency increases from $17 per acre at low risk aversion to $113 at the highest degree of risk. At the non-IPM factor proportions, the degree of allocative efficiency is uniformly higher, and declines less with risk aversion, as would be expected when pest risk is less in the earlier part of the season. At the low degree of risk aversion, the optimal input level is almost exactly equal to the observed value; it increases with the degree of risk aversion. The cost of allocative inefficiency rises from $8 per acre at the low degree of risk aversion to $39 at the high degree. Since average revenue per acre is about $1,300, these findings suggest that the allocative inefficiency costs could range from less than 1 percent of revenue to about 8.5 percent for the levels of risk aversion considered here.

The information in table 5–8 can be used to evaluate the likely effects of government regulations that restrict farmers' pesticide use. For

Table 5–8. Allocative Efficiency (AE) Indexes, Optimal Insecticide Input, Certainty Equivalents (CE) at the Optimum, and Cost of Allocative Inefficiency

	IPM				non-IPM			
Model	AE	Optimal input	Optimal CE	Ineff. cost	AE	Optimal input	Optimal CE	Ineff. cost
(A)	.975	1.786	49,444	1,304[a] 17	.988	.875	48,077	588 8
(B)	.950	1.829	50,059	2,740 36	.980	1.041	47,879	1,044 14
(C)	.910	1.847	51,309	5,603 75	.965	1.108	47,536	1,990 26
(D)	.879	1.852	52,598	8,444 113	.953	1.129	47,204	2,908 39

[a]The first number is the total inefficiency cost per 75-acre field, the second number is the cost per acre. AE is allocative efficiency as defined in equation (4.49). Actual input levels at group means were 1.44 for IPM, .88 for non-IPM. Models (A) to (D) represent increasing degrees of risk aversion, as defined in the text.

example, at the IPM factor proportions, it can be seen that if a farmer were actually producing with the optimal insecticide level implied by the model, and the government were to restrict insecticide use by about 25 percent, the resulting cost to the farmer would be a function of the degree of risk aversion. At the lowest degree of risk aversion, the cost would be only $17 per acre, but it would be $113 per acre if the farmer were very risk-averse. On the other hand, if pest risk were much less important (as in the case of fields planted early in the season), data in table 5–8 imply that the cost of restricting insecticide use would be much less. These findings demonstrate that policies which uniformly restrict pesticide use, without regard to the costs and benefits to farmers, impose disproportionately large costs on those growers who face the greatest pest risk and who are the most risk averse.

Policy makers might attempt to deal with the distributional inequities caused by uniform restrictions by making available an IPM technology that would allow farmers to substitute other inputs, such as knowledge and human labor, for pesticides. The allocative efficiency cost of restricting insecticide use would be made up, at least in part, by the welfare gain from the introduction of an IPM program that increased effectiveness of insecticide utilization. Moreover, as can be seen from tables 5–7 and 5–8, those who lose most from the restrictions (that is, farmers who face the most risk and are most risk averse) gain most from the introduction of the IPM technology. Thus *the implementation of an effective IPM program would tend to offset the negative welfare effects of pesticide restrictions in an equitable way.*

Turning now to the question of the bias in technological change induced by the IPM program, recall that the IPM program is designed to enable growers to determine more precisely, on the basis of observed pest population, whether or not they need to spray. If growers use exactly the same quantity and quality of insecticides each time they spray, the IPM program would tend, on average, to reduce insecticide use. This result can be likened to a substitution effect of the IPM technology on insecticide use. However, since the IPM program increases the likelihood that a given spray will be timed effectively, growers using that program have an incentive to spray with more effective (that is, higher quality) materials when they do spray. This result can be likened to an output effect of the IPM program on insecticide use. The technological bias of the IPM program in regard to insecticide use is determined by the sum of these two effects.

The technological bias is defined in terms of the change in the SMRTS between two inputs, as shown in chapter 3. Assuming that another input such as capital is unaffected by the IPM program, the bias between insecticide and that input depends on both the effect on the mean marginal products and on the marginal risk effects of the input. Since it has been found (table 5–2) that the mean marginal product of insecticides (quality-adjusted) is little affected by the IPM program, but that the program does increase the risk-reducing effects of pesticide, the evidence suggests that the IPM program biases the production technology toward increased use of insecticides, *measured in quality-constant units*. Whether this bias would increase the pounds of active ingredient applied depends on the way farmers choose to increase the effective units of insecticide.

The summary data in appendix A show that the IPM fields were treated on average with more insecticide, both in pounds of active ingredient and in quality-adjusted units. However, these means do not take into account the fact that the IPM fields were, on average, planted later in the season than the non-IPM fields, when pest problems are more severe. To investigate the differences in insecticide use, controlling for the effects of planting dates, the following regressions were estimated:

$$\underset{}{I} = \underset{(5.05)}{1.134} + \underset{(2.60)}{.807\,D2} + \underset{(2.80)}{1.107\,D3} + \underset{(1.29)}{.453\,IPM}, \quad R^2 = .139$$

$$IA = \underset{(6.728)}{1.523} + \underset{(1.260)}{.394\,D2} + \underset{(3.820)}{1.523\,D3} - \underset{(-1.504)}{.534\,IPM}, \quad R^2 = .137$$

where the numbers in parentheses are *t*-statistics and I = quality-adjusted insecticide per acre, IA = pounds of active ingredient per acre, $D2$ = midseason planting dummy, $D3$ = late-season planting dummy, and IPM = IPM dummy.

These results indicate, as expected, that the mid and late seasons were associated with higher usage of insecticide, whether measured in pounds of active ingredient or in quality-adjusted units. The results also show that after accounting for this timing effect, fields in the IPM program were treated with fewer pounds of active ingredient, but with larger quantities of quality-adjusted material. Thus the evidence does support the contention that the IPM program induces growers to substitute smaller amounts of higher-quality insecticides for lower-quality insecticides. The bias in the technology is therefore against insecticides in pounds of active ingredient, but toward insecticides measured in quality-adjusted terms.

Summary of Findings

In summary, the following evidence has been obtained from the case study of California processing-tomato production.

1. Accounting for sequential decision making and the resulting endogeneity of inputs was important. When the endogeneity was not accounted for in the econometric estimation of the technology, implausible results were obtained. This finding suggests that the IPM program did induce insecticide input to be endogenous.

2. The sample of producers was found to be both Arrow-Pratt and downside risk averse. The average risk premium was found to be about 11 percent of expected net returns. There was some evidence of decreasing absolute risk aversion, although the data indicated nearly constant absolute risk aversion at the population mean. Risk attitudes were not found to be heterogeneous, a result that probably reflects the homogeneity of the population, and the fact that the group of farmers in the sample were not randomly selected. Thus it remains an open question whether this sample was representative of the population of tomato growers in the Sacramento Valley.

3. Insecticides were found to be a marginally risk-reducing input. The marginal risk premium was found to be near zero for the non-IPM technology and more than 20 percent of expected returns for the IPM technology, indicating that the IPM program was highly effective in reducing damage risk.

4. The IPM program for tomatoes was found to yield welfare gains to producers. The program reduced mean damage significantly, and enhanced the risk-reducing effects of insecticides. The gains in technical efficiency provided by the program were greater later in the season, when pest damage was more severe. One implication of these findings is that the value of the IPM program to growers would be underestimated if the risk-reducing effects of the program were ignored, especially for growers producing late in the season.

5. The analysis of allocative efficiency showed that deviations from the optimal pesticide input level impose greater costs on those growers who face more pest risk and who are more risk averse.

6. The IPM technology was found to be biased against insecticides measured in pounds of active ingredient, but toward insecticides measured in quality-adjusted units. The growers in the case study were found to use smaller quantities of higher-quality insecticides, apparently because the IPM program increases the likelihood that insecticides would be used effectively.

One important implication of these findings for pesticide policy is that uniform restrictions on pesticide use are very inefficient in the sense that, to obtain a given reduction in use, they impose a high welfare cost on farmers. The inefficiency of uniform pollution standards has long been understood by environmental economists, who have proposed alternative policies to obtain more efficient pollution control. One suggested means of increasing efficiency would involve issuing tradeable permits for emissions (see Tietenberg, 1985): a polluting firm having a low cost for emissions reduction could sell permits to a polluter who had a high cost for emissions reduction, thus increasing the efficiency of the emissions restrictions.

It is unclear whether a similar scheme could be devised to deal with the agricultural pesticide problem. Because of the nonpoint nature of agricultural pesticide pollution, it is doubtful whether the agricultural pollution problem can be regulated through emissions. It is difficult, if not impossible, to identify and measure the emissions from each field or farm, so it would not be possible to know whether a farmer was in compliance with an emissions standard. One alternative might be to issue a fixed number of tradeable pesticide-use permits, an approach that would allow farmers who face more severe pest problems (that is, who have higher pollution control costs) to purchase the permits needed to apply more pesticides, while those having alternative control methods (low control costs) would have an incentive not to use pesticides.

Such a "pollution-trading" approach has some appeal from the point of view of economic efficiency, but it also entails both theoretical and practical problems. Theoretically, it is emissions that should be regulated, not inputs. Perhaps more important, it would be difficult to determine what pesticide-use level, as permitted by pesticide applications, would result in the desired level of emissions. In the regulation of agriculture, an additional problem is created by the large number of firms. It could be very costly to implement and enforce an emissions trading program in an industry with a large number of producers.

The foregoing analyses of the tomato IPM program, and the effects of allocative inefficiency on producer welfare, suggest that successful IPM programs provide an equitable way to reduce pesticide use. The IPM program's effects were equitable in the sense that they provided the greatest welfare gains to those producers who lost the most because of pesticide restrictions. The development of effective (that is, risk-reducing) IPM programs therefore could play a vital role in a public policy directed at the equitable reduction of pesticide externalities.

Notes on the Literature

The literature on pest management is vast and growing. There is also a smaller, yet substantial and growing literature on the relation of pest management to production risk. For reviews of the general literature, see McCarl (1981) and Osteen, Bradley, and Moffitt (1981);

see Carlson (1984) and Park (1985) for reviews of the risk literature. Although there appears to have been earlier recognition of the importance of risk in the use of pesticides, Carlson (1970) was one of the first to formally incorporate risk into a decision-theoretic model of pest management. Others to do so include Newton and Leuschner (1975) and Feder (1979). Empirical research on pest management methods has begun to take into account the risk attributes of production technologies and the risk attitudes of farm decision makers; see, for example, Liapis and Moffitt (1983) and Zavaleta and coauthors (1984).

Appendix A

Summary Statistics for the IPM and non-IPM Fields in the Sacramento Valley, Calif., Study

Variable	IPM	non-IPM
Fruit damage (percent)	.019 (.028)	.017 (.016)
Acreage	70.143 (31.660)	74.541 (44.122)
Fertilizer (lbs. *N*)	187.530 (50.486)	234.740 (40.560)
Water (acre-feet)	2.945 (.697)	3.249 (.743)
Insecticide (lbs. active ingredient)	1.759 (1.478)	1.455 (1.070)
Insecticide (quality-adjusted units)	2.494 (2.026)	1.645 (1.197)
Midseason dummy	.67	.49
Late-season dummy	.33	.00
Number of observations	21	85

Note: The data are sample means; standard deviations are given in parentheses.

Appendix B

An Alternative Model of the Firm's Objective Function

An alternative model of the firm's objective function is based on Machina's (1982) generalization of the expected utility maximization framework, discussed in chapter 4. In this model utility is defined to depend on net returns, as in the model in chapter 4, and also on the damage distribution directly, to reflect the fact that fruit damage has effects on the probability of a rejected load, as well as on future processing contracts. Thus the utility function is

$$U = U(\pi^*, f(D|\mu_1, \ldots, \mu_m)) \tag{B.1}$$

In this expression it has been assumed that the distribution of damage can be approximated in terms of the firm's m moments. Taking the expectation of utility, then, gives

$$EU = u[1 - \mu_1 - w^*x, \mu_1, \ldots, \mu_m] \tag{B.2}$$

Observe that expected utility in this formulation depends on the moments of net returns and the moments of the damage distribution; but since the moments of normalized net returns defined in model (5.4) equal the moments of damage, except for the mean, we obtain a similar expression, except that the mean of the damage distribution appears explicitly. While the conventional expected utility model and this more general model both imply that expected utility depends on the moments of the damage distribution, their interpretation is different. In particular, it is possible to link the parameters of model (5.4) to the Arrow-Pratt and downside concepts of risk aversion; that cannot be done for the model (B.2) because the reasoning related to the Taylor series approximation (discussed in chapter 4) cannot be used.

six

SUMMARY AND CONCLUSIONS

The purpose of this study is to provide an econometric framework for the measurement and analysis of the direct economic benefits that individual agricultural producers derive from the use of agricultural pesticides and from other pest management practices. The use of this framework has been illustrated with a case study of California processing-tomato production.

When economic efficiency criteria show there is a need for pesticide regulation because of pollution and other harmful side effects of pesticides, policy makers need to know the magnitude of the private costs that are imposed on producers by the regulations in order to balance the benefits of the regulations against their costs. Because a farmer's production decisions are based on ex ante expectations of prices and output, the conventional measures of the costs of pesticide restrictions, which are based on ex post yield losses, are not appropriate. It is ex ante welfare that is also relevant to the design and dissemination of technological alternatives to pesticides, such as integrated pest management programs. To evaluate producers' ex ante welfare, it is necessary to measure the stochastic production technology and producers' risk attitudes.

In chapter 2 it has been argued that the neoclassical theory of production is inappropriate for analysis of welfare and efficiency because it abstracts from two essential dimensions of pest management—sequential decision making and production uncertainty. Chapter 3 has shown how the neoclassical theory can be generalized to take these essential elements of agricultural production processes into account in evaluating producer efficiency and welfare. A theoretical model of firm behavior under uncertainty has two basic components, the stochastic technology and the behavioral model. Econometric methods for estimating the parameters of these components have been developed in chapter 4. By using a moment-based approximation to the stochastic technology, the distribution of risk attitudes can be estimated for the producer population. The estimated technology and risk attitudes can in turn be used to approximate the expected utility function and to evaluate the effects of pesticide policies on the efficiency and welfare of the producer population.

Chapter 5 has presented an application of this methodology to California processing-tomato production. This case study suggests that accounting for risk is very important in the analysis of pest management and the welfare effects of pesticide policies. The case study shows that pesticide restrictions have the greatest welfare effects on those producers who face the most production risk and who are the most risk averse. One implication of this finding is that uniform pesticide restrictions across producers facing different risks and possessing different risk attitudes are an inefficient and inequitable means of reducing pesticide use. The fact that the tomato IPM program has provided the greatest welfare gains to the very producers who lose the most from pesticide restrictions means that such IPM programs could play a role in an equitable pesticide policy.

Advocates of integrated pest management typically justify the need for IPM by arguing that there are often ecologically and economically sound alternatives to chemical-based pest management strategies. To the extent that they prove to be capable of generalization, the findings of this study provide an additional argument in favor of IPM as a tool for an equitable pesticide policy. IPM programs which substitute for the risk-reducing effects of pesticides offer the possibility of offsetting the welfare costs of pesticide regulations in an equitable manner, by providing the most benefits to those farmers who are most adversely affected by the regulations.

The methodological contribution of this study is the provision of a coherent framework in which decision making under uncertainty, and the essential stochastic, dynamic properties of agricultural production processes, can be accounted for in the analysis of producer efficiency and welfare. In addition, a feasible approach to the computation of efficiency and welfare measures, based on econometric estimates of the stochastic technology and producers' risk attitudes, has been proposed. This approach, grounded on the direct estimation of the primal technology (that is, on the distribution of output, revenue, or profits conditional on inputs) and on direct estimation of the producers' preferences, can be contrasted with the more conventional approach that is grounded on the indirect measurement of welfare as areas above or below supply or factor demand functions. This study has demonstrated that with readily available farm-level data, it is possible to make welfare evaluations using the proposed econometric strategy.

Certain further extensions and generalizations of the concepts and methods presented in this study are called for. First, it must be emphasized that the welfare measures developed here are valid at the individual producer level, or for a group of producers who face given product and input prices. However, as noted in chapter 2, aggregation to the industry level requires that the market equilibrium effects of policy or other parameter changes be taken into account. Thus to obtain a valid aggregate estimate of welfare change, it would be necessary to estimate market supply and demand shifts in product and factor markets in response to, say, the introduction of an IPM technology, and to make the resulting adjustments in equilibrium prices and quantities.

Second, further research should be devoted to the development of dynamic, stochastic production models which allow decision makers' risk attitudes to be taken into account. In this study, dynamic models with risk-neutral decision makers and static models with risk-averse decision makers have been discussed, and econometric procedures have been developed to account for the input endogeneity caused by sequential decision making, although the two types of models have not been explicitly integrated. In principle, the efficiency and welfare analyses discussed in this study can be readily adapted to the dynamic, risk-averse case; however, the introduction of explicit dynamics greatly complicates the analyses.

There are several possible alternative approaches to generalizing models to account explicitly for dynamics. One is to explicitly solve for the structural model (that is, the dynamic input demand functions), and to estimate the resulting structural equation model. The applicability of this approach seems to be limited to certain restrictive classes of functional forms. Another approach would be to derive and estimate models with their first-order conditions in implicit form. The methodology developed in chapter 4 is sufficiently general to allow for this approach, which appears to be a potentially fruitful avenue for further research. A third possible approach would be to use numerical simulation models. A variety of existing simulation models for pest management could be adapted to this purpose.

A third needed extension of the concepts and methods offered in this study is to incorporate several variable inputs in the analysis of the welfare effects of policies. In chapters 3 and 4, the theoretical and methodological developments allow for substitution possibilities between, say, pesticides and labor inputs, but the case study in chapter 5 treats only insecticides as variable. Since substitution possibilities often do exist, especially when one specific pesticide is restricted but other potential substitutes are not, it is important to further explore input substitution in the welfare analysis.

Two other extensions of the econometric methods proposed in this study deserve attention. One is the issue of the local and global curvature properties of the technology and the estimated objective function. For the efficiency and welfare analyses to be meaningful, the objective function must be at least locally concave in inputs. The estimated model's curvature properties depend on the combined properties of the technology (that is, the moment functions), the estimated risk attitude parameters, and the assumed form of the objective function (the expected utility function). In the application in chapter 5 the estimated model satisfies local concavity, but an estimated model may not be concave. When the estimated model fails the required curvature properties, there are two alternatives. One is to reformulate the model, on the assumption that the model that fails to satisfy the curvature properties is misspecified. The other is to devise methods to impose sufficient curvature properties in estimation. The latter approach involves computational difficulties, but is the only means of guaranteeing that the global curvature properties are satisfied.

Another extension of the econometric methods could address the existence of zero values for some inputs. For example, in the data used in the case study, some fields were not treated with insecticides. The presence of such zeros in the data pose potential bias problems in the risk attitude estimation procedures. If the insecticide demand functions were being estimated, there would be a "limited dependent variable problem" in estimation. In the approach developed in chapter 4, the risk attitude parameter estimates were obtained from the first-order conditions for expected utility maximization. Strictly speaking, these first-order conditions should satisfy the Kuhn-Tucker conditions for constrained maximization. This suggests that the statistical distributions associated with the first-order conditions should satisfy certain truncation properties which should be taken into account in the risk attitude estimation.

Future applied studies must be based on larger, longitudinal data sets. The data collection activities for the case study in chapter 5 showed that primary production data can be obtained from producers at a reasonable cost. The increasingly widespread use of personal computers by farm managers should reduce even further the costs of recording and retrieving primary production data. With longitudinal data, it would be possible to learn much more about the changes in technology over time, and about the distribution of risk attitudes across individuals and over time, and thus to more accurately evaluate changes in producer welfare.

REFERENCES

Anderson, J. R. 1973. "Sparse Data, Climatic Variability, and Yield Uncertainty in Response Analysis," *American Journal of Agricultural Economics* vol. 56, pp. 77–82.

______. 1979. "Perspectives on Models of Uncertain Decisions," in J. A. Roumasset, J. M. Boussard, and I. Singh, eds., *Risk, Uncertainty, and Agricultural Development* (New York, Agricultural Development Council).

______, J. L. Dillon, and B. Hardaker. 1977. *Agricultural Decision Analysis* (Ames, Iowa, Iowa State University Press).

Antle, J. M. 1983a. "Sequential Decision Making in Production Models," *American Journal of Agricultural Economics* vol. 65, pp. 282–290.

______. 1983b. "Testing the Stochastic Structure of Production: A Flexible Moment-Based Approach," *Journal of Business and Economic Statistics* vol. 1, pp. 192–201.

______. 1984. "Incorporating Risk in Production Analysis," *American Journal of Agricultural Economics* vol. 65, pp. 1099–1106.

______. 1985. "Technical and Allocative Efficiency Under Uncertainty," Renewable Resources Division Discussion Paper RR85-07, Resources for the Future, Washington, D.C.

______. 1986. "Aggregation, Expectations, and the Explanation of Technological Change," *Journal of Econometrics* vol. 33, pp. 213–236.

______. 1987. "Econometric Estimation of Producers' Risk Attitudes," *American Journal of Agricultural Economics* vol. 69, pp. 509–522.

______. 1988. "Dynamics, Causality, and Agricultural Productivity," in S. M. Capalbo and J. M. Antle, eds., *Agricultural Productivity: Measurement and Explanation* (Washington, D.C., Resources for the Future).

______, and S. M. Capalbo. 1986. "Pesticides and Public Policy," in T. Phipps, P. Crosson, and K. Price, eds., *Agriculture and the Environment* (Washington, D.C., National Center for Food and Agricultural Policy, Resources for the Future).

______, and ______. 1988. "An Introduction to Recent Developments in Production Theory and Productivity Measurement," in S. M. Capalbo and J. M. Antle, eds., *Agricultural Productivity: Measurement and Explanation* (Washington, D.C., Resources for the Future).

______, and W. J. Goodger. 1984. "Measuring Stochastic Technology: The Case of Tulare Milk Production," *American Journal of Agricultural Economics* vol. 66, pp. 342–350.

______, and S. A. Hatchett. 1986. "Dynamic Input Decisions in Econometric

Production Models," *American Journal of Agricultural Economics* vol. 68, pp. 939–949.

———, and A. Havenner. 1983. "Formulating and Estimating Dynamic Stochastic Production Models," Working Paper no. 83-8, Department of Agricultural Economics, University of California, Davis.

———, and S. K. Park. 1985. "Econometric Analysis of Pest Management Technology." Manuscript, Department of Agricultural Economics, University of California, Davis.

Aoki, M. 1967. *Optimization of Stochastic Systems* (New York, Academic Press).

Arrow, K. J. 1971. *Essays in the Theory of Risk-Bearing* (Amsterdam, North-Holland Publishing).

Berndt, E. R., and L. R. Christensen. 1973. "The Internal Structure of Functional Relationships: Separability, Substitution, and Aggregation," *Review of Economic Studies* vol. 40, pp. 403–410.

———, and D. O. Wood. 1979. "Engineering and Economic Interpretations of Energy-Capital Complementarity," *American Economic Review* pp. 342–344.

———, C. J. Morrison, and G. C. Watkins. 1983. "Dynamic Models of Energy Demand: An Assessment and Comparison," in E. R. Berndt and B. C. Field, eds., *Modeling and Measuring Natural Resource Substitution* (Cambridge, Mass., MIT Press).

Binswanger, H. P. 1980. "Attitudes Towards Risk: Experimental Measurement in Rural India," *American Journal of Agricultural Economics* vol. 62, pp. 395–507.

———. 1982. "Empirical Estimation and Use of Risk Preferences: Discussion," *American Journal of Agricultural Economics* vol. 64, pp. 391–393.

———, and D. A. Sillers. 1983. "Risk Aversion and Credit Constraints in Farmer Decision Making: A Reinterpretation," *Journal of Development Studies* vol. 20, pp. 5–21.

———, D. Jha, T. Balaramaiah, and D. Sillers. 1982. "The Impact of Attitudes Toward Risk on Agricultural Decisions in Rural India," Economics Program Progress Report no. 42, International Crop Research Institute for the Semi-Arid Tropics (ICRISAT).

Brookshire, D. S., M. A. Thayler, J. Tschirhart, and W. E. Schulze. 1985. "A Test of the Expected Utility Model: Evidence from Earthquake Risks," *Journal of Political Economy* vol. 93, pp. 369–389.

Carlson, G. A. 1970. "A Decision Theoretic Approach to Crop Disease Prediction and Control," *American Journal of Agricultural Economics* vol. 52, no. 2, pp. 216–222.

———. 1977. "Long-Run Productivity of Pesticides," *American Journal of Agricultural Economics* vol. 59, no. 3, pp. 543–548.

———. 1984. "Risk Reducing Inputs Related to Agricultural Pests," in *Risk Analysis of Agricultural Firms: Concepts, Information, Requirements and Policy Issues*, Proceedings of Regional Research Project S-180, Department of Agricultural Economics, University of Illinois, Urbana.

Chavas, J.-P., and R. D. Pope. 1981. "Welfare Measures of Production Activities under Risk Aversion," *Southern Economic Journal* vol. 48, pp. 187–196.

Coombs, C., R. Dawes, and A. Tversky. 1970. *Mathematical Psychology* (Englewood Cliffs, N.J., Prentice-Hall).

Coulston, F. 1979. *Regulatory Aspects of Carcinogenesis and Food Additives: The Delaney Clause* (New York, Academic Press).

Council for Agricultural Science and Technology. 1980. *Social and Economic Impacts of Restricting Pesticide Use in Agriculture,* Report no. 86 (Ames, Iowa).

Day, R. H. 1965. "Probability Distributions of Field Crops," *Journal of Farm Economics* vol. 47, pp. 713–741.

de Janvry, A. 1972. "Fertilization under Risk," *American Journal of Agricultural Economics* vol. 54, pp. 1–10.

Diamond, P., and M. Rothschild. 1978. *Uncertainty in Economics* (New York, Academic Press).

Edwards, W., and A. Tversky. 1967. *Decision Making* (Harmondsworth, England, Penguin Books).

Engle, R. F. 1984. "Wald, Likelihood Ratio, and Lagrange Multiplier Tests in Econometrics," in Z. Griliches and M. D. Intrilligator, eds., *Handbook of Econometrics,* vol. 2 (Amsterdam, North-Holland Publishing).

Epstein, L. G. 1985. "Decreasing Risk Aversion and Mean-Variance Analysis," *Econometrica* vol. 53, pp. 945–961.

———, and A. J. Yatchew. 1985. "The Empirical Determination of Technology and Expectations: A Simplified Procedure," *Journal of Econometrics* vol. 27, pp. 235–258.

Farrell, M. J. 1957. "The Measurement of Productive Efficiency," *Journal of the Royal Statistical Society,* series A, vol. 140, pp. 253–281.

Feder, G. 1979. "Pesticides, Information, and Pest Management under Uncertainty," *American Journal of Agricultural Economics* vol. 61, pp. 97–103.

Ferguson, C. 1971. *The Neoclassical Theory of Production and Distribution* (Cambridge, Cambridge University Press).

Fishburn, P. C. 1980. "Stochastic Dominance and Moments of Distributions," *Mathematics of Operations Research* vol. 5, pp. 94–100.

Flint, M. L., and R. van den Bosch. 1981. *Introduction to Integrated Pest Management* (New York, Plenum Press).

Forsund, F. R., C. A. K. Lovell, and P. Schmidt. 1980. "A Survey of Frontier Production Functions and of Their Relationship to Efficiency Measurement," *Journal of Econometrics* vol. 13, pp. 5–25.

Freeman, A. M. 1979. *The Benefits of Environmental Improvement* (Baltimore, The Johns Hopkins University Press).

Freund, J. E. 1972. *Mathematical Statistics* (2d ed., London, Prentice-Hall).

Gallant, A. R., and D. W. Jorgenson. 1979. "Statistical Inference for a System of Simultaneous, Implicit Equations in the Context of Maximum Likelihood Estimation," *Journal of Econometrics* vol. 11, pp. 275–302.

Grisley, W., and E. D. Kellogg. 1983. "Farmers' Subjective Probabilities in Northern Thailand: An Elicitation Analysis," *American Journal of Agricultural Economics* vol. 65, pp. 74–82.

Hansen, L. P. 1982. "Large Sample Properties of Generalized Method of Moments Estimators," *Econometrica* vol. 50, pp. 1029–1054.

———, and T. J. Sargent. 1980. "Formulating and Estimating Dynamic Linear Rational Expectations Models," *Journal of Economic Dynamics and Control* vol. 2, pp. 7–46.

Hatchett, S. A. 1985. "Dynamic Input Decisions: An Econometric Analysis of Crop Response to Irrigation" (Ph.D. dissertation, University of California, Davis).

Hazell, P. B. 1982. "Application of Risk Preference Estimates in Firm-Household and Agricultural Sector Models," *American Journal of Agricultural Economics* vol. 64, pp. 384–390.

Hazilla, M., and R. J. Kopp. 1986. "Testing for Separable Functional Structure Using Temporary Equilibrium Models," *Journal of Econometrics* vol. 33, nos. 1, 2, pp. 119–142.

Headley, J. C. 1968. "Estimating the Productivity of Agricultural Pesticides," *American Journal of Agricultural Economics* vol. 50, pp. 13–23.

———. 1972. "Defining the Economic Threshold," in *Pest Control Strategies for the Future* (Washington, D.C., National Academy of Sciences).

———, and J. N. Lewis. 1967. *The Pesticide Problem: An Economic Approach to Public Policy* (Baltimore, The Johns Hopkins University Press).

Herdt, R. W., and A. M. Mandac. 1981. "Modern Technology and Economic Efficiency of Philippine Rice Farmers," *Economic Development and Cultural Change* vol. 29, pp. 375–400.

Hicks, J. R. 1963. *The Theory of Wages* (New York, St. Martins Press).

Hildreth, C., and J. P. Houck. 1968. "Some Estimators for a Linear Model with Random Coefficients," *Journal of the American Statistical Association* vol. 63, pp. 584–595.

Hsiao, C. 1983. "Identification," in Z. Griliches and M. D. Intrilligator, eds., *Handbook of Econometrics*, vol. 1 (Amsterdam, North-Holland Publishing).

Jorgenson, D. W. 1963. "Capital Theory and Investment Behavior," *American Economic Review* vol. 53, pp. 247–259.

Just, R. E. 1975. "Risk Aversion Under Profit Maximization," *American Journal of Agricultural Economics* vol. 57, pp. 347–452.

Just, R. E., and R. D. Pope, 1978. "Stochastic Specification of Production Functions and Economic Implications," *Journal of Econometrics* vol. 7, pp. 67–86.

———, and ———. 1979a. "On the Relationship of Input Decisions and Risk," in J. A. Roumasset, J. Boussard, and I. Singh, eds., *Risk, Uncertainty, and Agricultural Development* (New York, Agricultural Development Council).

———, and ———. 1979b. "Production Function Estimation and Related Risk Considerations," *American Journal of Agricultural Economics* vol. 61, pp. 277–284.

———, and D. Zilberman. 1984. "The Role of Micro-Risk Behavior in Aggregate Relationships," Giannini Foundation of Agricultural Economics, Berkeley, Calif., mimeographed.

———, D. L. Hueth, and A. Schmitz. 1982. *Applied Welfare Economics and Public Policy* (Englewood Cliffs, N.J., Prentice-Hall).

Kelejian, H. H. 1971. "Two-Stage Least Squares and Econometric Systems Linear in Parameters but Nonlinear in the Endogenous Variables," *Journal of the American Statistical Association* vol. 66, pp. 373–374.

Kendall, M., and A. Stuart. 1977. *The Advanced Theory of Statistics*, vol. 1 (4th ed., New York, Macmillan).

Kendrick, D. 1981. "Control Theory with Applications to Economics," in K. J. Arrow and M. D. Intrilligator, eds., *Handbook of Mathematical Economics*, vol. 1 (Amsterdam, North-Holland Publishing).

Khaneman, D., and A. Tversky. 1979. "Prospect Theory: An Analysis of Decisions Under Risk," *Econometrica* vol. 47, pp. 263–291.

Knight, F. 1921. *Risk, Uncertainty, and Profit* (Boston, Houghton-Mifflin).

Kopp, R. J. 1981. "The Measurement of Productive Efficiency: A Reconsideration," *Quarterly Journal of Economics* vol. 95, pp. 477–503.

Leamer, E. E. 1978. *Specification Searches* (New York, Wiley).

Liapis, P. S., and L. J. Moffitt. 1983. "Economic Analysis of Cotton Integrated Pest Management Strategies," *Southern Journal of Agricultural Economics*, pp. 97–102.

Lippman, S. A., and J. J. McCall. 1981. "The Economics of Uncertainty: Selected Topics and Probabilistic Methods," in K. J. Arrow and M. D. Intrilligator, eds., *Handbook of Mathematical Economics*, vol. 1 (Amsterdam, North-Holland Publishing).

Lockheed, M. E., D. T. Jamison, and L. J. Lau. 1980. "Farmer Education and Farmer Efficiency: A Survey," *Economic Development and Cultural Change* vol. 29, no. 1, pp. 37–76.

Lucas, R. E. 1967. "Adjustment Costs and the Theory of Supply," *Journal of Political Economy* vol. 75, pp. 321–334.

Luce, R. D., and P. Suppes. 1965. "Preference, Utility, and Subjective Probability," in R. D. Luce, R. Bush, and E. Galanter, eds., *Handbook of Mathematical Psychology*, vol. 3 (New York, Wiley).

Machina, M. J. 1982. "'Expected Utility' Analysis Without the Independence Axiom," *Econometrica* vol. 50, pp. 277–323.

Marschak, J., and W. J. Andrews. 1944. "Random Simultaneous Equations and the Theory of Production," *Econometrica* vol. 12, pp. 143–205.

May, L. L. 1984. "Dynamic Risk / Benefit Analysis in Pesticide Regulation: The Case of Toxaphene" (M.A. thesis, University of California, Riverside).

McCarl, B. A. 1981. "Economics of Integrated Pest Management: An Interpretive Review of the Literature." Oregon State University, Agricultural Experiment Station Special Report 636.

McFadden, D. 1983. "Econometric Models of Probabilistic Choice," in C. F. Manski and D. McFadden, eds., *Structural Analysis of Discrete Data with Econometric Applications* (Cambridge, Mass., MIT Press).

McKenzie, G. 1983. *Measuring Economic Welfare: New Methods* (Cambridge, Cambridge University Press).

Menezes, C., G. Geiss, and J. Tressler. 1980. "Increasing Downside Risk," *American Economic Review* vol. 70, pp. 921–932.

Moscardi, E., and A. de Janvry. 1977. "Attitudes Toward Risk Among Peasants: An Econometric Approach," *American Journal of Agricultural Economics* vol. 59, pp. 710–716.

Mundlak, Y., and I. Hoch. 1965. "Consequences of Alternative Specifications in Estimation of Cobb-Douglas Production Functions," *Econometrica* vol. 33, pp. 814–828.

Murphy, K. M., and R. H. Topel. 1985. "Estimation and Inference in Two-Step Econometric Models," *Journal of Business and Economics Statistics* vol. 3, pp. 370–379.

Murtaugh, B. A., and M. A. Saunders. 1977. "MINOS User's Guide," Technical Report 77-9, Department of Operations Research, Stanford University.

Nerlove, M. 1958. *The Dynamics of Supply: Estimation of Farmers' Response to Price* (Baltimore, The Johns Hopkins University Press).

Newbery, D. M. G., and J. E. Stiglitz. 1981. *The Theory of Commodity Price Stabilization* (Oxford, England, Clarendon Press).

Newton, C. M., and W. A. Leuschner. 1975. "Recognition of Risk and Utility in Pest Management Decisions," *Bulletin of the Entomological Society of America* vol. 21, pp. 169–171.

Osteen, C. D., E. B. Bradley, and L. F. Moffitt. 1981. "The Economics of Agricultural Pest Control," Bibliographies and Literature of Agriculture Series no. 14, U.S. Department of Agriculture, Economics and Statistics Service.

Park, S. K. 1985. "Econometric Measurement of Pest Management Technology: Risk and the Economics of a Worm Monitoring Program for Processing Tomatoes in the Sacramento Valley" (Ph.D. dissertation, University of California, Davis).

Parke, W. R. 1986. "Pseudo Maximum Likelihood Estimation: The Asymptotic Distribution," *Annals of Statistics* vol. 14, pp. 355–357.

Pratt, J. W. 1964. "Risk Aversion in the Small and in the Large," *Econometrica* vol. 32, pp. 122–136.

Pope, R. D. 1982. "Empirical Estimation and Use of Risk Preference: An Appraisal of Estimation Methods That Use Actual Economic Decisions," *American Journal of Agricultural Economics* vol. 64, pp. 376–383.

———, and R. A. Kramer. 1979. "Production Uncertainty and the Factor Demands for the Competitive Firm," *Southern Economic Journal* vol. 60, pp. 425–435.

———, and J.-P. Chavas. 1985. "Producer Surplus and Risk," *Quarterly Journal of Economics* vol. 100, supplement, pp. 853–869.

———, ———, and R. E. Just. 1983. "Economic Welfare Evaluations for Producers Under Uncertainty," *American Journal of Agricultural Economics* vol. 65, pp. 98–107.

Rao, C. R. 1973. *Linear Statistical Inference and Its Application* (New York, John Wiley and Sons).

Ross, S. A. 1981. "Some Stronger Measures of Risk Aversion in the Small and the Large with Applications," *Econometrica* vol. 49, pp. 621–638.

Roumasset, J. A. 1976. *Rice and Risk: Decision Making Among Low Income Farmers* (Amsterdam, North-Holland Publishing).

Rudin, W. 1964. *Principles of Mathematical Analysis* (2d ed., New York, McGraw-Hill).

Schoemaker, P. J. H. 1982. "The Expected Utility Model: Its Variants, Purposes, Evidence and Limitations," *Journal of Economic Literature* vol. 20, pp. 529–563.

Schultz, T. W. 1975. "The Value of the Ability to Deal with Disequilibria," *Journal of Economic Literature* vol. 13, pp. 827–846.

Stigler, G. J., and G. S. Becker. 1977. "De Gustibus non est Disputandum," *American Economic Review* vol. 67, pp. 76–90.

Stoker, T. M. 1984. "Completeness, Distribution Restrictions, and the Form of Aggregate Functions," *Econometrica* vol. 52, pp. 887–907.

Taylor, C. R. 1980. "The Nature of Benefits and Costs of Use of Pest Control Methods," *American Journal of Agricultural Economics* vol. 62, pp. 1007–1011.

______. 1984. "A Flexible Method for Empirically Estimating Probability Functions," *Western Journal of Agricultural Economics* vol. 9, pp. 66–75.

Tietenberg, T. H. 1985. *Emissions Trading* (Washington, D.C., Resources for the Future).

Timmer, C. P. 1971. "Using a Probabilistic Frontier Function to Measure Technical Efficiency," *Journal of Political Economy* vol. 79, pp. 776–794.

University of California. 1982. *Integrated Pest Management for Tomatoes* (Berkeley, University of California Press).

von Neuman, J., and O. Morgenstern. 1947. *Theory of Games and Economic Behavior* (Princeton, Princeton University Press).

Weakley, C., F. Zalom, and L. T. Wilson. 1986. "Worm Sampling Methods for Processing Tomatoes," *California Tomato Grower* vol. 29, no. 4.

White, H. 1984. *Asymptotic Theory for Econometricians* (New York, Academic Press).

Willig, R. D. 1976. "Consumer's Surplus Without Apology," *American Economic Review* vol. 66, pp. 589–597.

Zavaleta, L. R., B. Eleveld, M. Kogan, L. Wax, D. Kuhlman, and S. M. Lim. 1984. "Income and Risk Associated with Various Pest Management Levels, Tillage Systems, and Crop Rotations: An Analysis of Experimental Data," Agricultural Economics Report no. 191, Department of Agricultural Economics, University of Illinois, Urbana.

Zellner. A., J. Kmenta, and J. Dreze. 1966. "Specification and Estimation of Cobb-Douglas Production Function Models," *Econometrica* vol. 34, pp. 784–795.